The Great Implosion

T. E. Mitchell

Copyright © 2017 by T. E. Mitchell.

Library of Congress Control Number: 2017908588
ISBN: Hardcover   978-1-5434-2694-6
              Softcover   978-1-5434-2696-0
              eBook   978-1-5434-2695-3

All rights reserved. No part of this book may be reproduced or transmitted in any form or by any means, electronic or mechanical, including photocopying, recording, or by any information storage and retrieval system, without permission in writing from the copyright owner.

Any people depicted in stock imagery provided by Thinkstock are models, and such images are being used for illustrative purposes only.
Certain stock imagery © Thinkstock.

Print information available on the last page.

Rev. date: 06/05/2017

To order additional copies of this book, contact:
Xlibris
1-888-795-4274
www.Xlibris.com
Orders@Xlibris.com
762742

Table of Content

Chapter One . . . Creation of the Universe — Page 7
    What is Real? — Page 8
    First Law of the Universe — Page 10
    First Division of Energy — Page 11
    Second Division of Energy — Page 13
    Second Law of the Universe — Page 16
    Third Division of Energy — Page 18
    Third Law of the Universe — Page 22
    Fourth Law of the Universe — Page 23
    Fifth Law of the Universe — Page 26
    Fourth Division of Energy — Page 27
    Reviewing the Color Force — Page 32

Chapter Two . . . Space and Time — Page 40
    Power of Force — Page 41
    Expansion of Force — Page 43
    Revision of Relativity — Page 44
    Velocity through Space — Page 45
    Depths of Space — Page 47
    Electromagnetic Radiation — Page 48
    Size of the Universe — Page 49
    Travel through Space — Page 53
    Puffs of Star Dust — Page 55
    Infinities — Page 56
    Power of C — Page 58
    Power of h — Page 61

Chapter Three . . . Unification of Force — Page 63
    Strong Force — Page 65
    Color Force — Page 66

Charge Force	Page 69
Nuclear Force	Page 70
Weak Force	Page 77
Gravity Effect	Page 78
Mass Force	Page 83
Chapter Four . . . Equations	Page 86
Calculations	Page 86
Chapter Five . . . Basic Particles 101	Page 92
Quark Family	Page 92
Anti-Quark Family	Page 93
W Family	Page 94
Proton and Neutron	Page 97
Negative Proton	Page 100
Anti-Particle Family	Page 101
Gluons	Page 102

Forward

What you are about to read is a simple unifying theory based on existing laws, principals, and equations within the scientific community. I write this book not because I want too but because I must. The scientific community has reached a stalemate in its research when it comes to how the universe was created and how nature functions. In this book, I present some new concepts and ideas which hopefully will lead the way to new breakthroughs in research and applications. I present these new theories so the average person can understand how simple the universe functions.

Chapter One . . . Creation of the Universe

In the beginning, the universe had a radius of 6.62×10^{-27} centimeters by our standard of measure today using centimeters, grams, and seconds. It had a diameter of 13.24×10^{-27} centimeters or 1.324×10^{-26} centimeters. Try to imagine a perfect sphere of force equal to a force value of $F = 1.359 \times 10^{47}$ we will call the Singular Universal Force or SUF. The "true nature" of the SUF is expansion. There was no mass, no particles, not even space itself. Time had yet to begin. There was only a perfect sphere of force.

This sphere of force began to expand at the velocity of C or 3×10^{10} centimeters per second creating space and time. The value of the force diminished to create space and time until it reached a value of one and a radius of 9×10^{20} centimeters. At a radius of 9×10^{20} centimeters there was a great explosion that created all the mass particles in the universe. Equal amounts of particles and anti-particles were created around the edge of this perfect sphere of force called the Event Horizon.

These two spheres of particles that existed at the event horizon formed the more complex particles that exist today. One must understand how the universe created particles with mass to understand how all the anti-particles had their properties reversed to become particles.

Reversing the properties of anti-particles cannot be performed today because the conditions that existed at the event horizon no longer exist. From a classical point of view, it was a one shot or one event happening within an area of space one hundred million times smaller than the size of a proton.

So, how did the universe create particles from an expanding sphere of force? One must understand what is real.

"What is real?"

A century ago it was believed particles were composed of a changeable material. This belief gave rise to the concept that the universe was made up of real solid particles. Unfortunately, double slit experiments showed that these particles also behaved like waves. Light was believed to be packets of energy but they also behaved like waves when it came to the double slit experiments. Electrons were believed to be real solid particles composed of an electrical substance but they also behaved like waves in the double slit experiment. These experiments gave rise to a real dilemma if one believed particles were composed up of a real substance in space. The dilemma was resolved by describing particles as wave particles.

Unfortunately, we must dismiss both beliefs. Particles are not composed up of a real substance you can hold in your hand nor are they simply waves in space. Particles are fields of force in space and where the fields of force end you find your particle region. However, there is nothing inside these particle regions. Particles are simply "bubbles" of energy within expanding space. We incorrectly think a particle having mass takes up a certain amount of space within space. A particle having mass displaces space within space.

Figure 1

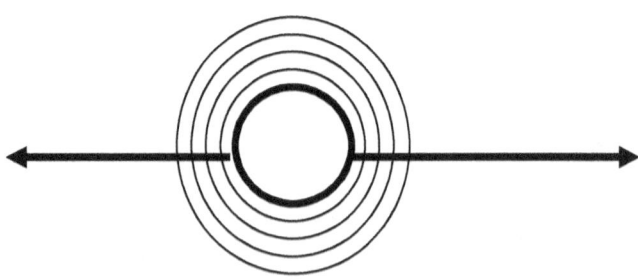

The closer you get to a particle region the stronger these fields of force become. The amount of energy conserved within these fields of force determines the actual size of the particle within space. Where the field of force ends one could say the particle region begins. The fields of force curves space around itself creating the illusion there is a particle within the fields of force. It is important to understand there are no actual particles in the center of these fields.

When we first built accelerators strong enough to create particles, we discovered equal amounts of anti-particles were also created. When we create an electron we also create a positron. The positron meets up with another electron and they annihilate each other. Two photons and two neutrinos are the results of the annihilation. It was then we should have realized that particles are not composed up of any real material.

Imagine this for a moment. You are an electron and you have been around since the beginning of the universe. You have encountered countless interactions with other electrons. Along comes a new positron just created by a bunch of scientists and your fields begin to cancel each other out. Your mass is converted back into energy and space closes in on you. Finally, you simply disappear from the universe in a puff of photonic energy and a neutrino that carries off your existence.

There is no way you could be a piece of real solid material. Nor are you simply a wave in space. You are fields of force with properties that define you. Your rest mass is only energy conserved within your fields of force that resist change. There is nothing inside of you. All that matters, is what is outside of you. Your fields of force have mass that curves space around you creating the illusion you are a particle when you are nothing but fields of force with properties.

So how did the universe create these particles that are not real?

We must first have a complete understanding of force, energy, and mass. Force is the ability to create change. Energy is the amount of change being created. Mass is the resistance to change. Understand these principals and you will understand how the universe created itself. I must use classical examples you can visualize so you will understand the differences between force, energy, and mass.

We must also keep in mind a very important premise when we explore the inwards of the universe, "The Universe is the way it is because it is the only way it can be." We cannot insert a square peg into a round hole.

First Law of the Universe

"For every action, there is an equal and opposite reaction."

These particles and anti-particles stepped down from the velocity of C and entered this expanding sphere of force taking on mass as their velocity diminished from the value of C. There were particles expanding outward within this expanding sphere of force and equal amounts of anti-particles compressing inward within this expanding sphere of force. Based on the first law of the universe, you cannot create a particle without creating an anti-particle. This will become evident when we look at all the particles our accelerators create and when we look at the basic changes of particles within the nucleus of our atoms.

Try to imagine two perfect spheres of particles within an expanding sphere of force. The outer sphere of particles expanded outward with the expanding sphere of force while the inner sphere of anti-particles compressed inward within this sphere of force creating a great implosion of anti-particles as shown in Figure 2 below.

Figure Two

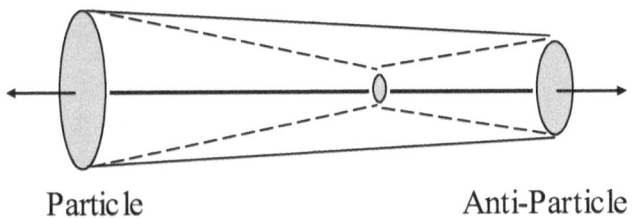

Particle   Anti-Particle

These anti-particles possessed liked charges. One can view this perfect sphere of anti-particles compressing inward on itself as existing in a reversed state of time. Having liked charges, the compression became so great their momentum was brought to a stand still and their properties were reversed by emitting anti-neutrinos. This sphere of anti-particles compressing inward emerged from this great implosion as particles with positive momentum and time and expanded outward with the other sphere of particles. All the anti-particles created were reversed and became particles by emitting anti-neutrinos. This all happened within an instant of time when all the matter of the universe was created along the event horizon with a radius of 9×10^{20} centimeters.

First Division of Energy

"The Strong Force emerged converting kinetic Energy into Mass"

I want you to image a single dot that represents force as shown below in Figure 3. If we grab this dot on opposite sides and stretch it, the dot becomes a string. It takes energy to stretch this dot into a string. Here we have force (dot) diminishing into energy (string) as shown in Figure 4 below.

Figure 3

●

Force

For those of you who believe in the String Theory, I am going to make your day but only for a brief moment in time. Our dot of force diminishes into a string of energy. Our string of energy stretches into one dimension and that is depth. At the event horizon, there are only two opposite directions our strings of energy can stretch into: (1) inward into existing space and (2) outward into expanding space.

Figure 4

When we stretch our dot into a string of energy we see energy is dividing into two opposite directions. Here we have our first law of action and reaction in play. From this point on, I will use arrows to represent these two directions. We will keep absolute track of all the arrows as the number of opposing arrows must always be equal, representing the division of energy. Energy will divide and change four times to create particles having mass. For every action, there must be an opposite and equal reaction.

The energy conserved within our string of energy takes on mass. Each end of our string can represent a virtual particle taking on mass. A virtual particle is a particle in transition, not yet complete. Mass curves space giving us structure and the effect of gravity. Mass is also the resistance to change. Force diminishing into energy is our action and energy converting into mass is our reaction, curving space and time. Once the energy force within Mass equals one and our diminishing Singular Universal Force reaches zero, our dot of force vanishes and stops diminishing into energy. This is what happened after the universe expanded out from 6.62×10^{-27} centimeters to a radius of 9×10^{20} centimeters. At a radius of 9×10^{20} centimeters our SUF value had diminished to a value of one. Strings of energy penetrated this expanse of space with a force of one. We can view each end of these strings of energy as a virtual Higgs particle forming a Higgs field having a radius of 6.62×10^{-27} centimeters and a diameter of 13.24×10^{-27} centimeters in space as shown in Figure 5 below.

Figure 5

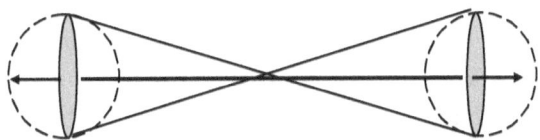

Second Division of Energy

"Displacement of Space creates Fields of Force"

The curvature and displacement of space by particles taking on mass created fields of force within space, giving the universe opposite fields of "charged" particles. For this to happen there had to be a second division of energy. We can represent this division of energy by twisting each end of our string of energy in opposite directions. We represent this division of energy with two opposing vertical arrows as shown in Figure 6 below. As much energy went into twisting our strings of energy as went into stretching our dot of force into strings of energy.

Figure 6

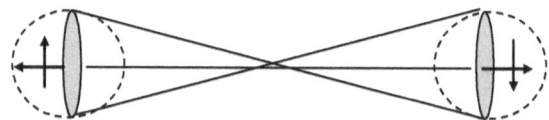

We know that mass curves space and time giving us structure and the effect of gravity. Our SUF or Singular Universal Force diminished creating strings of energy which is our action. At each end of our strings of energy, the energy converted to mass which is our reaction. As our SUF diminished from a value of one to zero our Mass increased from a value of zero to a value of one, giving us action and reaction. Our Mass stopped our SUF from diminishing any further than it did. Force is our ability to cause change and Mass is our resistance to change. Energy is the amount of change that occurred between the two opposing forces. Take note that I have declared Mass as a force and not gravity. We will learn later that gravity is an effect and is not a force like the forces of SUF, Strong, Charge, Mass, and Color. From this point on I will refer to the force that converted the string of energy into mass as the "Strong Force". Do not mistake it as the Strong Force that holds the proton and neutrons within the nucleus of an atom. The force that holds protons and neutrons together will be called the Nuclear Force in this book.

The "true nature" of the Strong Force is the conversion of kinetic energy or velocity to mass which we view as attraction. The Strong Force is only felt over a very short distance. It is the Strong force that keeps energy in the form of rest mass.

Without particles having mass, the universe would not exist. The Singular Universal Force or SUF is an expanding force we view as "repulsion" while the Strong Force is a contracting force we view as "attraction".

There is a simple equation we can use to determine at what radius it took the Strong Force to stop the SUF from diminishing any further and the converting of its energy into mass. First, we must convert Force into units of energy:

$$E = v \times h$$

The amount of energy within light is equal to the frequency or "v" times Planck's constant. Therefore, all we do is to divide the amount of the SUF Force we have by Planck's constant:

$$F = 1/h = 1/6.62 \times 10^{-27} = 1.51 \times 10^{26} \text{ units Planck's Constants}$$

Now that we have the number of energy units in Planck's Constants we can use my equation in physics to determine the gravitational constant of the universe when our SUF force value was equal to one:

$$G_v = (hv)^2 = ((6.62 \times 10^{-27}) \times (7.55 \times 10^{25}))^2 = 2.498 \times 10^{-1}$$

Since the total energy was divided equally at each end of our Higgs particles their energy units in Planck's constants were 7.55×10^{25} units of energy. We can now fill in the values in the equations below:

$$E = h v = (6.62 \times 10^{-27}) \times (7.55 \times 10^{25}) = .49981$$

$$M = E/C^2 = .49981 / 9 \times 10^{20} = .0555 \times 10^{-20} = 5.55 \times 10^{-22}$$

$$F = G_v \times M_1 \times M_2 / R^2 \quad \text{and} \quad R = \sqrt{G_v \times M_1 \times M_2 / F}$$

$$R = \sqrt{(2.498 \times 10^{-1}) \times (5.55 \times 10^{-22}) \times (5.55 \times 10^{-22})} = 2.77 \times 10^{-22}$$

All the energy within the two ends would have been converted to mass at a radius of 2.77×10^{-22} centimeters reducing our SUF of one to a value of zero. The attractive Strong Force within our two mass particles would have brought our two particles to a stop relative to each other at a radius of 2.77×10^{-22} centimeters and a diameter of 5.55×10^{-22} centimeters. This distance is critical to our two particles escaping the "Strong" force. The amount of time it took was 385,000 Planck seconds or 2.55×10^{-22} of a second.

At the event horizon 2.5×10^{85} strings of energy created a particle and anti-particle within a tubular radius of 6.62×10^{-27} centimeters. The curvature of the event horizon was one centimeter for every 30,000,000,000 centimeters (30 billion centimeters). Our system of virtual particles exists within 13.24×10^{-27} centimeters of space in width which means there are 7.55×10^{25} virtual particles outward and 7.55×10^{25} virtual anti-particles inward per centimeter of length, (75,000,000,000,000,000,000,000,000 = 75 million, million, million particles). Our 2.5×10^{85} outward virtual particles have momentum in the direction of expansion while our 2.5×10^{85} inward virtual anti-particles have momentum inward.

Second Law of the Universe

"What you begin with you must end up with."

Our second law is a conservation law that requires what you begin with you must end up with after you have an action with an equal and opposite reaction. We are going to represent action and reaction with opposing arrows. Our dot of SUF force having a value of one has stretched into a string of energy and the value of our SUF has diminished to a value of zero. We can view each end of this string of energy as a virtual particle. To give these virtual particles opposing fields of force, we must twist each end of our string of energy in opposite directions. Twisting these ends requires energy so we have a second division of energy to create virtual particles. We represent this division of energy with two opposing vertical arrows as shown in Figure 6 above. As much energy went into twisting our string of energy as went into stretching our string of energy.

We know that energy has a certain amount of gravity associated with it. An incredible amount of energy was created when our force diminished to a value of zero. This energy was conserved at each end of our string by the "Strong" force. Conserved energy is Mass. This incredible amount of energy created an enormous amount of mass and this enormous amount of mass curved space giving us structure in the form of fields. The attractive nature of the "Strong" force grew so great that our virtual particles could no longer separate and our string of energy could no longer be stretched and twisted. Our SUF Force could no longer diminish. We have two opposite virtual particles at each end of our string of energy.

We must stop and realize there is only one force in the universe but this force takes on many "faces" or different values because our force is diminishing, energy is being created and is dividing, and particles with mass are being created. The forces of Strong, Charge, Mass, and Color are only different "faces" of the Singular Universe Force or SUF.

Take note when I write of a particle having mass, I am referring to a "bubble" of energy displacing space within space creating spherical fields of force caused by the displacement of space within space and not to an actual material particle existing within space. Our minds can visualize a particle far more easily than it can visualize a spherical field of force displacing space within space.

Now, if we let go of each end of our string of energy as shown in Figure 6 above, our string of energy will untwist and stretch back to a dot of force. There is only one way to turn our virtual particles into real field particles.

So how did our universe turn virtual field particles into real or independent particles?

The Third Division of Energy

"Particles with Mass Create Paired but Opposite Particles with Color"

Energy had to divide again for a third time and each particle had to create a pair of opposite particles. One of these opposite particles would be emitted to push our original particles out pass the grip of the "Strong" force which only acts over a radius of 2.77×10^{-22} centimeters or a diameter of 5.55×10^{-22} centimeters. These opposite particles are sacrificed to collapse the "Strong" force fields and the strings of energy that bind them as shown in Figure 7 and Figure 8 below.

Figure 7

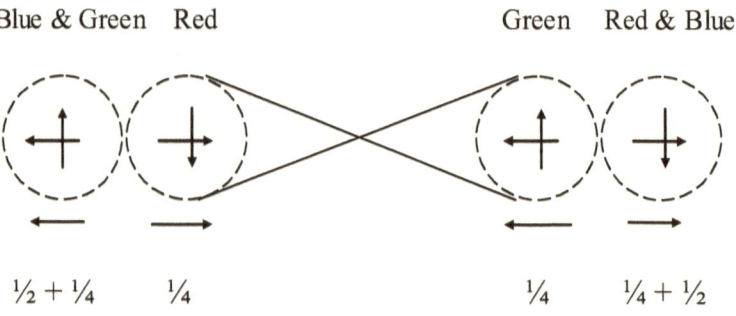

With this third division of energy the "Strong" force had to diminish and spread its force over a greater distance to decompress the new particles creating the weaker "Color" force which we represent with three colors: (1) blue, (2) green, and (3) red. Whereas the "Strong" force is only an attractive force, the "Color" is a combination of the SUF and the "Strong" force. Color repels and attracts by the means of emitting and absorbing colored gluons which is our fourth division of energy but first we must first understand how our original particles created our paired but opposite particles. It is the fundamental principal of the universe when force diminishes, energy is created, and energy divides creating equal but opposite particles.

When the force within a particle with mass diminishes, energy is created from its mass and the energy divides creating two equal but opposite particles with mass-energy: (1) particle and (2) anti-particle. The mass-energy values "flowing" within these two new particles are different from each other and different from the particle whose force is diminishing, giving us three particles with different mass-energy levels or "Color".

Two of these new particles will attract each other and separate from the other particles as shown in Figure 8 below. Their opposite particles will join with the original two particles. I do not show a particle within a particle because they are joined together having different sizes (See Figure 10) and the images would be too confusing. Instead I show two horizontal arrows within a circle joined together by their "charge" arrow as shown in Figure 8 below.

Another way of visualizing the outer particles in Figure 8 below consisting of two joined particles is to imagine blowing bubbles with soapy water through a circular bubble chamber device. We have all done it. Sometimes when we blow bubbles we create two bubbles joined together with one bubble being larger than the other. The smaller bubble rides on the surface of the larger bubble. If one bubble breaks they both break. As shown in Figure 8 below their like "charges" combine to equal the original "charge" value of one.

Figure 8

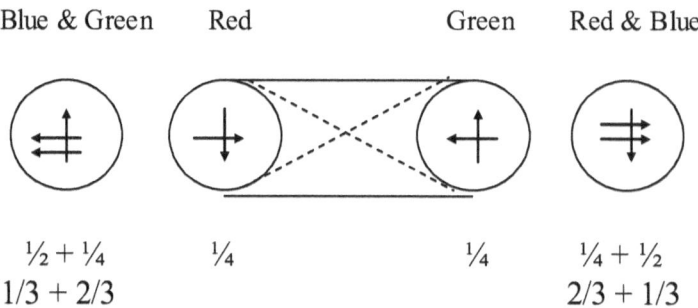

Blue & Green	Red	Green	Red & Blue
½ + ¼	¼	¼	¼ + ½
1/3 + 2/3			2/3 + 1/3

The string of energy contracts with the two new particles having unlike "charge" and "color" leaving two real but unstable outer particles. These two new inner particles will combine and form a third neutral particle as shown in Figure 9 below.

Figure 9

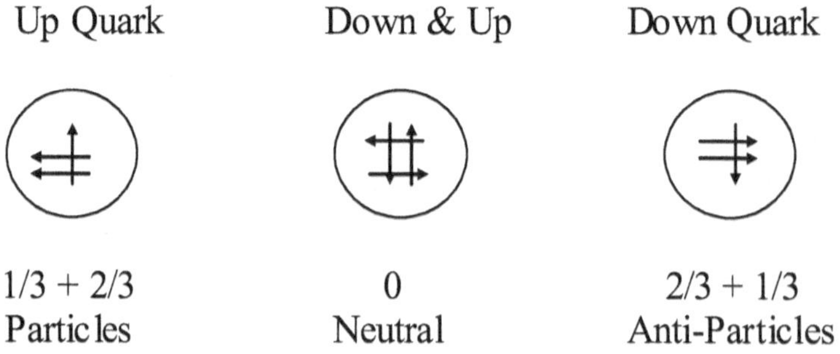

Up Quark	Down & Up	Down Quark
1/3 + 2/3	0	2/3 + 1/3
Particles	Neutral	Anti-Particles

The size or diameter of the particles determines the "charge" and field strength of the particle and the size or diameter of the particles is determined by the amount of rest mass-energy within the particles. So, the smaller particles having the less rest mass within them will have the greater charge because it has a smaller diameter within space. The smaller its diameter gets to 6.62 x 10⁻²⁷ centimeters the greater its field strength becomes as shown in Figure 10 below.

The combined "charge" of the two outer particles cannot be greater than or less than the charge of the original particle. The original particles now have half the mass-energy they had so they are smaller in diameter. The particles they created and combined with have a quarter of the mass-energy of the original particle. The original particle has twice the mass-energy of the new particle it combined with so it takes on one third (1/3) of the "charge" because it displaces a greater amount of space within space as shown in Figure 10 below.

Figure 10

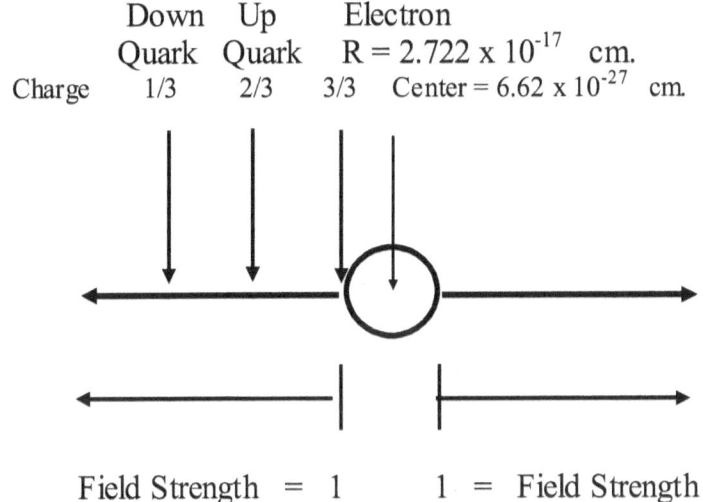

The particle it combined with has half the mass-energy so it takes on two thirds (2/3) of the "charge". The relationship between the ½ and the ¼ is 1/3 and 2/3 which is equal to 3/3 or three thirds of the "charge" value.

The two new inner particles attract each other and combine forming an unstable neutral particle as shown in Figure 9 above. The other two outer independent particles are also unstable, unstable meaning they are joined together with more than one particle or field. These particles are unstable because they represent two unequal divisions of energy taking on mass that curves space and time around them. We will call these particles "quarks". We will label these quarks based on their direction of momentum within space and time, particle and anti-particle. These quarks have "charge" which we will label the up quark as positive and the down quark as negative. The particle in the center is labeled neutral because it will never take on any direction of momentum and it will decay into a neutrino and anti-neutrino. The unlike "charge" and "color" of the two particles annihilating each other are canceled out and the neutrinos will take on ½ spin either up or down in the direction of the charge it had when it was created as shown in Figure 8 above.

Third Law of the Universe

"All interactions within the Universe must be reversible or they will not occur."

The neutral particle will decay into two neutrinos with one of them being an anti-neutrino as show in Figure 11 below. The neutrino is the "lock out" particles of the universe. Without neutrinos, the universe would not have been able to reverse all the anti-particles to particles and the universe would have annihilated itself before it had a chance to begin. Without neutrinos, quarks would not have been able to form protons and neutrons. Without neutrinos protons and neutrons would not have been able to form heavier elements within the universe. Without neutrinos, stars would not have fusion and there would be no light shining down on earth to create life. Without neutrinos, life in the universe would not exist.

Figure 11

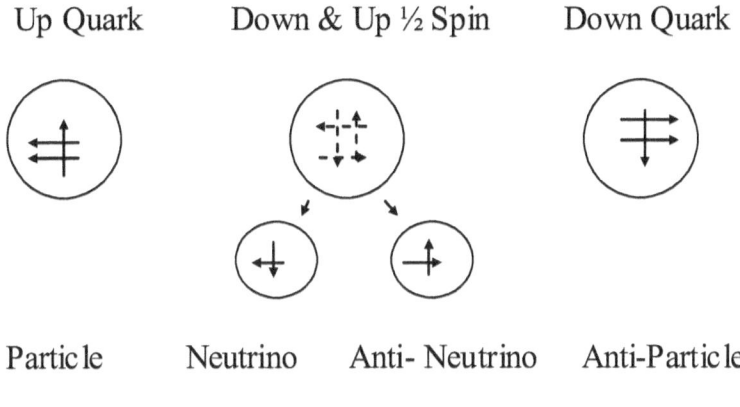

Fourth Law of the Universe

"Conservation of Information and Number must be preserved."

    The neutrino and anti-neutrino carry off the interaction information of the two particles that were created and annihilated each other to rid the universe of its strings of energy and push our two unstable outer particles outside the barrier of the Strong force, having a diameter of 5.55×10^{-22} centimeters. This area of space where the Strong force is active is 100 million times smaller than the size of a proton or neutron today.

    The sphere of up quarks is expanding out into expanding space. The sphere of down anti-quarks is compressing inward into existing space. This sphere of down anti-quarks has the same charge and for a split moment in time their momentum came to a standstill. The two anti-quarks emitted two anti-neutrinos as show in Figure 13 below reversing the properties and the momentum of the anti-quarks. Our anti-quarks become quarks based on the direction of their "charge".

Figure 12

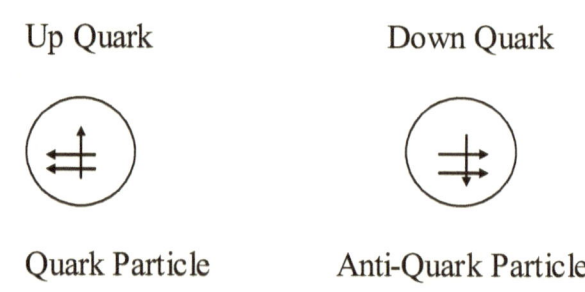

We must view these anti-quarks as two particles joined together by their mutual "charge" and existing as one particle with different "strong" force values. The third division of energy came with a third "face" of force we call the "color force" which is white as shown in Figure 13 below.

Figure 13

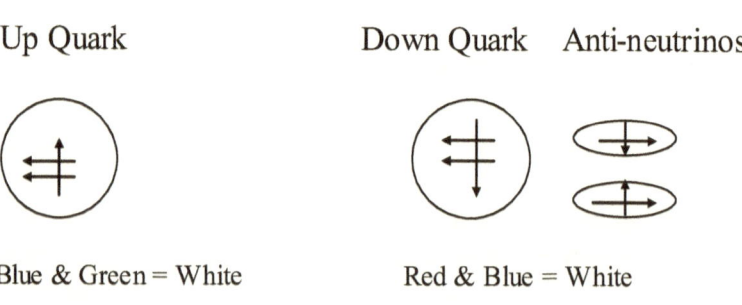

But first we must understand how the down anti-quarks became down quarks by flipping its horizon arrows and emitting two anti-neutrinos with an Up and Down ½ Spin. This process is easier understood by viewing the vertical "charge" arrow as one down arrow when the original particle combined with the new particle combining their "charge". The two particles "charge" had to equal a value of one. When the down anti-quarks came to a standstill, its two horizontal arrows and one vertical arrow diminished back to dots of energy. These dots of energy divided again as show in the Figure 14 below.

Our down anti-quark "A" can be viewed as in "B" and due to the compression of the sphere of down anti-quarks their two horizontal arrows and one vertical arrow diminish back into dots of energy as shown in "C" and the first horizontal arrow and the vertical arrow divides again as in "D" and "E" becomes temporarily a virtual up quark by emitting an anti-neutrino with ½ Down Spin "F". "E" becomes a virtual up quark and the vertical up arrow then collapses or diminishes back to a dot of energy with the second horizontal arrow.

The second horizontal and vertical dots divide again creating "G" and emits an anti-neutrino with ½ Up Spin "I" and the "G" becomes "H" or a down quark. The vertical arrow flips up and down or twice while the two horizontal arrows flip once. The total numbers of arrows are equal and the charge within the universe is balanced including the spin of the anti-neutrinos.

Figure 14

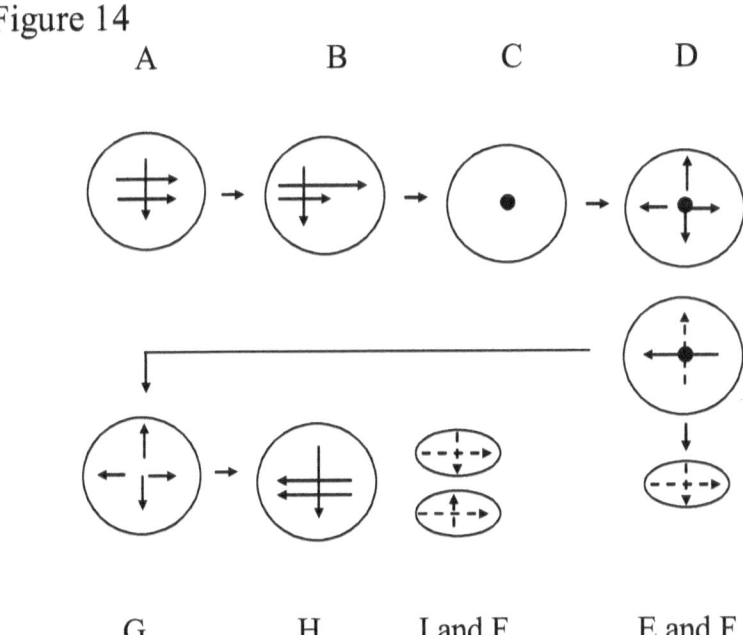

An easier way of viewing this change is to flip the upper horizontal arrow up and around 180 degrees and the anti-quark becomes a quark by emitting an anti-neutrino with a ½ down spin. Flip the lower horizontal arrow down and around 180 degrees and the anti-quark becomes a quark by emitting an anti-neutrino with a ½ up spin. Whenever a horizontal arrow flips its direction 180 degrees a neutrino or an anti-neutrino is emitted.

Fifth Law of the Universe

"The Pauli Exclusion Principle forbids two particles from having the same properties within the same space and time."

The Pauli Exclusion Principle forbids two or more particles from having the same properties within the same space and time. In other words, there must be a difference in their properties or the two particles will be one in the same. Our two quarks above having two horizon arrows with different mass-energy can be represented with different colors of the "color" force. There are three colors in the "color" force: red, green, and blue. We can have two colors of "blue" because our quarks have different "charges" within separate particles. Our two unstable quarks cannot divide into equal parts having the same charge and color because they have different mass-energy levels.

Figure 15

The "Strong" and "Charge" force diminished giving us down quarks with 1/3 "charge", up quarks with 2/3 "charge", and free electrons and positrons quarks with 3/3 "charge".

Fourth Division of Energy

"Unstable Particles Decay and Energy Divides creating Gluons"

"Particle Decay, Combine, and Emit"

The fourth division of energy created gluons that are the force carrying particles of the "color" force. But first we must look at how the three quantum states of the "charge" force created our basic proton and neutron particles by decaying, combining, and emitting free electron quarks (while we ignore the "color" force in the figures). Each vertical "charge" arrows we have used in our classical examples must now be looked upon as three 1/3 "charge" arrows, up and down.

Our two unstable quarks in Figure 15 above decay by mutual attraction forming four quarks as shown in Figure 16 below. The Up quarks decay and becomes a positive +2/3 up quark and a positive +1/3 up quark. The Down quarks decay and becomes a negative -2/3 down quark and a negative -1/3 down quark.

Our positive +1/3 Up quark combines with our negative -2/3 Down quark. The +1/3 positive charge cancels out -1/3 negative charge of our -2/3 Down quark making it a negative -1/3 Down quark as shown in Figure 16 below. We now only have three quarks, a positive +2/3 Up quark and two negative -1/3 down quarks making our system of quarks a neutron. However, one of our negative -1/3 down quarks is unstable with two horizontal lines as shown in Figure 16 below.

Here is our weak force in action. Our unstable negative -1/3 Down quark emits a -W particle which is a free electron quark that emits an anti-neutrino. For this to happen our negative -1/3 quark must become a positive +2/3 Up quark making our system of quarks a proton with a free negative electron.

We now have two positive +2/3 Up quarks and one negative -1/3 Down quark giving our system of quarks a positive "charge" of one. Our electron has a negative "charge" of one. "Charge" in the universe is balanced out and the conservation law for number of particles created returns to four. The actual process is a lot more complicated since we have not included the color of our quarks in the "color force" which is the diminished "Strong" force.

A proton can become a neutron by emitting a +W particle which is a positron that emits a neutrino. The positron annihilates an electron and the "charge" balance is restored. To understand the weak force one must understand the difference between a meson and W particle which is a free electron or positron quark.

We must take note the positive +1/3 Up quark and the negative -2/3 Down quark that combined to form an unstable negative -1/3 Down quark in Figure 16 no longer exist within our system of quarks once our unstable negative -1/3 Down quark emits a -W particle or electron plus an anti-neutrino and becomes a stable positive +2/3 Up quarks.

After our -W particle emits its anti-neutrino our electron is forever stable and free. We only have two types of stable quarks within our system of quarks that form either a proton or neutron and they are a positive +2/3 Up quark and a negative -1/3 Down quark. The universe made itself simpler.

Within a proton or neutron, the universe is still creating anti-particles with its particles but none of the anti-particles will ever survive. The universe has a unique way of getting rid of all the anti-particles it creates by emitting anti-neutrinos.

We are now left with a system of three very stable quarks that form our basic proton with an electron or a neutron without an electron. There are only two types of quarks that exist, a positive +2/3 Up quark and a negative -1/3 Down quark.

Figure 16

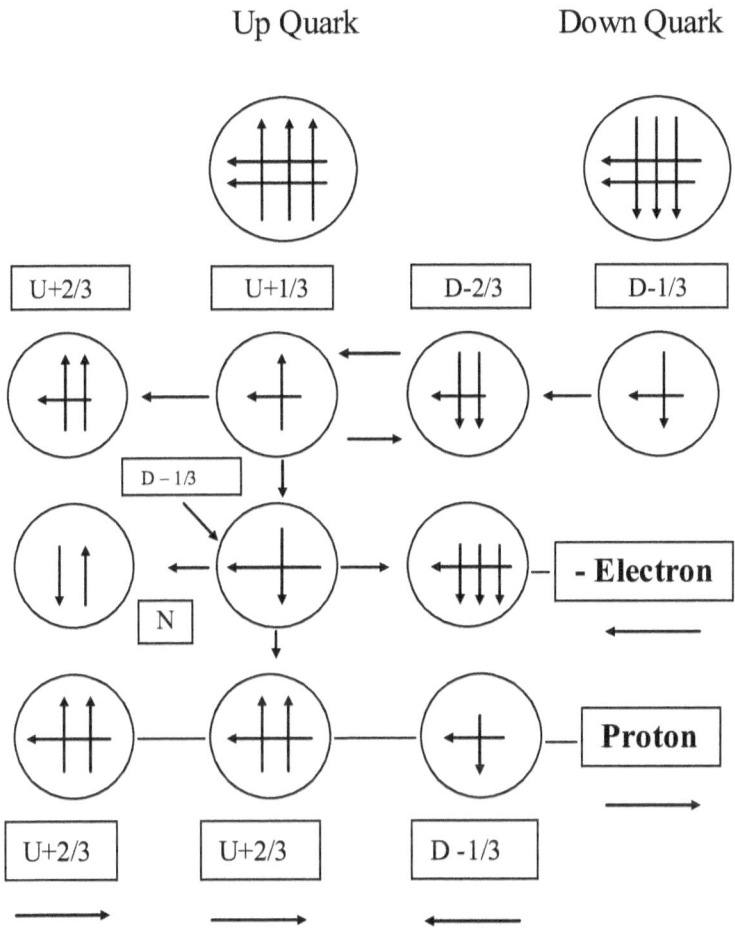

Two Up quarks and a Down quark form our basic proton and one Up quark and two Down quarks form our basic neutron. To form heavier nuclei or elements we need both protons and neutrons. Protons must be able to change into neutrons and neutrons must be able to change back into protons.

This process is called the "Weak" force but it is the same process that created all the particles with mass-energy in the universe.

We shall look at the "Weak" force and how a positive +2/3 Up quark can change itself into a negative -1/3 Down quark. The Up quark simply creates a Higgs particle and anti-particle as shown in Figure 17 below.

Figure 17

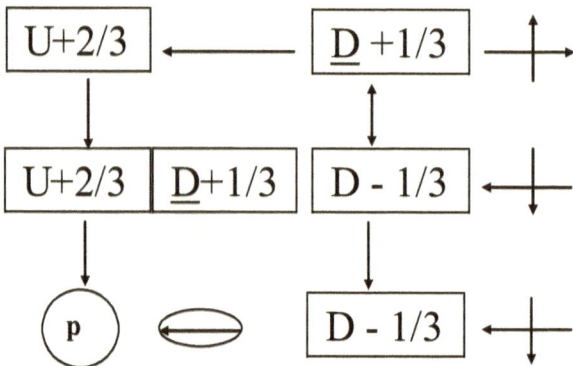

The Up quark creates a negative -1/3 Down-quark and a positive +1/3 Down anti-quark. When we create a particle, we must create an equal and opposite anti-particle. Our Up quark joins with our anti-Down quark and is emitted as a +W particle or "p" positron leaving in its place our negative -1/3 Down-quark within our system of three quarks. Our +W particle consist of a particle and anti-particle joined together. It's total "charge" adds up to a positive one. Here we must look at the direction of our energy "flow" of our two particles. Our positive +1/3 anti-down quark has twice the mass-energy flowing to the right than our positive +2/3 Up quark which "flows" to the left. The "flow" of mass-energy in each particle is canceled out or annihilated by 1/3 leaving 1/3 of mass-energy flowing to the right and equal to the amount of energy to make a positron (See Figure 10) which has less mass-energy in it than an Up and Down quark.

Our +W particle is now a positron with energy flowing to the right. Its horizontal arrow pointing to the right making it an anti-particle. The annihilated mass-energy from both particles is transferred to the neutrino it emits. The arrow of the neutrino is pointing to the left. The neutrino is what is left of our annihilated +2/3 Up quark and carries off its information of the quark's existence in the universe. Its arrow points to the left with a ½ Up spin representing the positive charge of the annihilated, Up quark. It is a neutrino with an Up spin.

The positron will annihilate the first electron in its path. The two particles will annihilate each other. Two photons, a neutrino, and an anti-neutrino will be what is left from the two particles annihilating each other.

Now we shall look at how a negative -1/3 Down quark can change itself into a positive +2/3 Up quark. The Down quark simply creates a Higgs particle and anti-particle as shown in Figure 18 below.

Figure 18

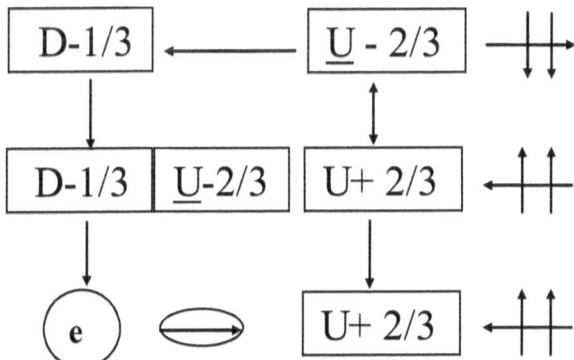

The Down quark creates a positive +2/3 Up-quark and a negative -2/3 Up anti-quark. When we create a particle, we must create an equal and opposite anti-particle. Our Down quark joins with our anti-Up quark and is emitted as a -W particle leaving in its place our positive +2/3 Up-quark within our system of three quarks. Our -W particle consist of a particle and anti-particle joined together. It's total "charge" adds up to a negative one. Here we must look at the direction of our energy "flow" of our two particles. Our negative -1/3 Down quark has twice the mass-energy flowing to the left than our negative -2/3 anti-up quark which "flows" to the right. The "flow" of mass-energy in each particle is canceled out or annihilated by 1/3 leaving 1/3 of mass-energy flowing to the left and equal to the amount of energy to make an electron (See Figure 10) which has less mass-energy in it than an Up and Down quark.

Our -W particle is now an electron with energy flow to the left. Its horizontal arrow pointing to the left making it a particle. The annihilated mass-energy from both particles is transferred to the anti-neutrino it emits. The arrow of the anti-neutrino is pointing to the right. The anti-neutrino is what is left of our annihilated -2/3 anti-up quark and carries off its information of the quark's existence in the universe. Its arrow points to the right with a ½ Down spin representing the negative charge of the annihilated Anti-Up quark. It is an anti-neutrino with a Down spin. The electron will become a separate part of the three quarks system and the "charge" becomes balanced.

Reviewing the Color Force

When energy divided a third time another "face" of the Singular Universal Force was created with the third division of energy. The "Strong Force" emerged with the first division of energy converting the strings of energy to virtual Higgs particles with mass. It was the equal reaction to the SUF or Singular Universal Force diminishing from a value one to zero. The "Charge Force" emerged at the same time the "Strong Force" did with the second division of energy creating fields of force within space around our virtual bubbles of energy having mass. The "Color Force" emerged with the third division of energy when four new particles were created having less energy than their original particles to break the strings of energy and setting our virtual particles free. Two of the four new particles annihilated each other leaving two unstable quarks that broke apart and created our three quarks system with a free quark electron.

The Pauli Exclusion Principal forbids two particles within a system from having the same properties and energy levels or they will occupy the same space and time. This problem becomes apparent when we look at the quarks within our proton and neutron structures. A proton consists of one negative -1/3 Down-quark and two positive +2/3 Up-quarks as shown in Figure 19 below.

Figure 19

There had to be a difference between the two Up quarks or they could not exist within the proton as separate quarks. Scientists showed a difference by stating one Up quark had an Up Spin and the other Up quark had a Down Spin. This statement worked until the Omega Particle showed up. The Omega particle consist of three Up quarks as shown in Figure 20 below.

Figure 20

The Omega proved an unknown force had to exist between the three Up quarks for the Omega particle to be created in our accelerators. The need for a three "color force" was also showing up in the creation of other particles. However, the universe does not care about our needs. The "color" force emerged with the third division of energy.

We have three states and three particles where the "charge" force have different strengths: (1) Down-quark with a charge of 1/3, (2) Up-quark with a charge of 2/3, and the free quark (electron or positron) with a charge of 3/3. These particles' charges are determined by the size or displacement of space within space and their size is determined by the amount of rest mass within the particles.

There should be no surprise to discover there are three colors to the "color force". It is a surprise the "color" force is a variable force with no set strengths unlike the "charge force" which have quantum "charge" strengths. However, the strength of the "color force" cannot add up to a strength greater than one or the strength of our "Strong Force".

The Color Force is an attractive and repulsive force existing between quarks within protons and neutrons that have three variant force values we define as Red, Green, and Blue. However, the color force started out with the color of "White" before it fractured to hold three quarks within the protons and neutrons. It was the Strong Force that converted all the energy within the strings of energy to mass that created the first Higgs Particles. The strength of the color force is a variant limited to the power of one which was the strength of SUF when mass particles were created at the edge of the event horizon. The equation below determines the force value of the color force:

$$\text{Color Force} = \frac{R^2}{h}$$

The three quarks within a proton or neutron can only be separated from each other by a center radius of 8.136×10^{-14} of a centimeter. At that radius, the color force of the quarks is equivalent to a value of one. With a color force value of one, all energy supplied to the quarks to separate them any further is converted to mass. It is equivalent to the Strong Force.

$$C_F = \frac{R^2}{h} = \frac{(8.136 \times 10^{-14})^2}{6.62 \times 10^{-27}} = \frac{6.62 \times 10^{-27}}{6.62 \times 10^{-27}} = 1$$

The radius of a proton and neutron just happens to be 8.136×10^{-14} of a centimeter. The color force of the quarks within a proton structure limits the size of a proton and neutron to a radius of 8.136×10^{-14} of a centimeter. The diameter of a proton and neutron is 1.6×10^{-13} of a centimeter or less.

The quarks in our proton and neutrons change their color by emitting gluons between them. There are six types of gluons quarks can emit to keep the quarks apart from each other and escaping from the proton or neutron structure they are a part.

Take note that the strength of the different "colors" of the "color force" is not written in stone. Each "color" can have a variant of different strengths as long as the total of their colors adds to the "color force" or "White". The strength of each color is dependent on the distance between two or three colored quarks. A "blue" and "red" can add up to white. A "blue" and "green" can add up to white. A "red", "green", and "blue" can add up to white. Keep in mind that the weaker "colored force" is only the "Strong Force" that converted kinetic energy into mass.

Fourth Division of Energy

"Unstable Particles Decay and Energy Divides creating Gluons"

"Energy Divides Creating Gluons"

The fourth division of energy created gluons that are the force carrying particles of the color force. The fundamental principal of creation is when force diminishes, energy is created and divides into a particle and an anti-particle. The "true nature" of the color force is the same as the "Strong" force from which it comes, attraction.

The only difference between the "Strong" Force and the "Color" Force is the distance over which they act. The "Strong" Force acts over a smaller radius of 2.77×10^{-22} of a centimeter while the "Color" Force acts over a radius of 8.136×10^{-14} of a centimeter.

The "true nature" of the "Strong" Force is the conversion of kinetic energy or velocity to mass which we view as attraction. The "Strong" Force is only felt over a very short distance. It is the "Strong" force that keeps energy in the form of rest mass.

Each of our three quarks in our protons and neutrons have a different color representing the "color" force. Unlike colors attract each other. Like colors repel each other. When two of our three quarks having unlike colors attract each other their "color" force diminishes in strength. When a force diminishes, energy is created, and energy divides creating a particle and an anti-particle. When it comes to the "color" force a gluon is created. There are six different types of gluons that can be created. A gluon is a composition particle composed up of a particle and anti-particle created at the same time and still joined together unlike the Higgs' particle. They have no mass because they are not separated. A gluon is a force carrying particle of the color force like the photon is the force carrying particle of the "charge" force.

Figure 21

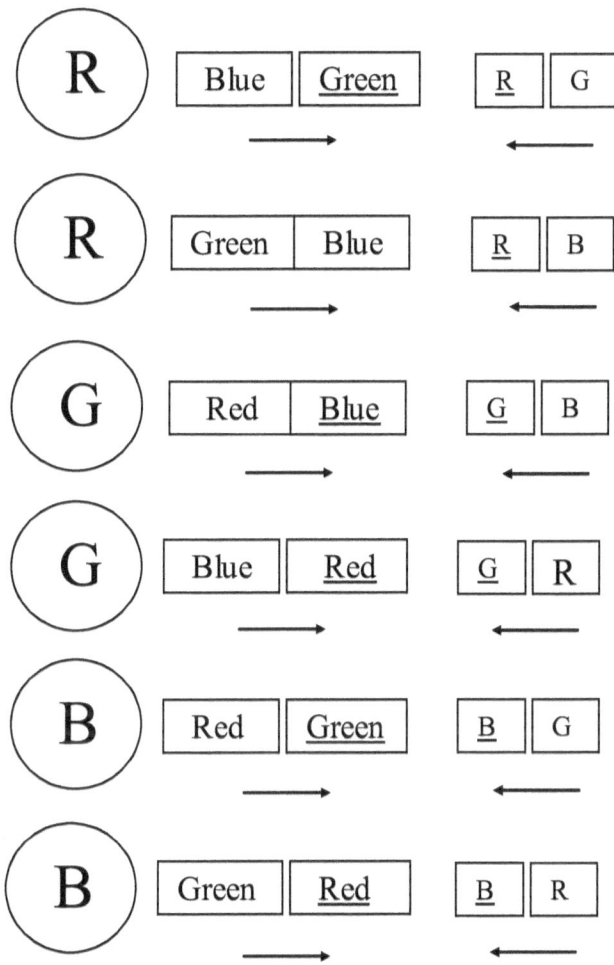

As shown in Figure 21 a red quark can only emit or absorb a blue-green or a green-blue gluon. A green quark can only emit or absorb a red-blue or a blue-red gluon. A blue quark can only emit or absorb a red-green or a green-red gluon.

The fourth division of energy came when the Up and Down quarks separated and created gluons to keep them separated. Additional energy went into the division and separation of the quarks creating "Spin" and "Parity". Rotation of the quarks had to occur to bring about their final separation. Each quark took an opposite direction of "Spin" either up or down. Unlike the rotation of the earth around its axis once, the quarks not only see their rotation but the rotation from the quark it separated from. In a classical sense, our quarks rotate twice or 720 degrees to get back to where it started. This is required so reversal of events can occur. No particle is truly independent but is linked all the way back to the beginning of the universe.

Parity assures you see the same view as you rotate around a quark. There is no front and back to our quarks. You only see only the same view of a quark. Spin and parity were the finally ingredients to forming a proton or neutron in space.

I have shown you how the universe created the first particles with "mass" and "charge" and rid itself of all the "strings" and "anti-matter" in the universe by dividing their energy and emitting anti-neutrinos during a great implosion of all the anti-particles. The two remaining particles were unstable and divided into four particles weakening their "charge" and "strong" forces. With the emission of an electron, the neutrons changed into protons, the star dust that formed all the stars and galaxies in the universe. This all happened within a radius of 2.77×10^{-22} centimeters, the area of space where the "Strong Force" is active and a radius of 8.136×10^{-14} of a centimeter, where the "Color Force" is active. The "Strong Force" diminished into the "Color Force" active over a distance 100 million times larger than the area of the "Strong Force". This all happened within 2.55×10^{-22} of a second or 385,000 Planck seconds at the edge of the event horizon with a radius of 9 x

10^{20} centimeters. Around the edge of this spherical event horizon 2.5 x 10^{85} protons and neutrons were created by stepping down from the speed of light and exploded within the expanding space of the universe. All the rest is basic particle physics 101.

All the anti-particles in the universe were changed to particles by the anti-particles emitting two anti-neutrinos. All the laws of the universe were conserved and none were violated to form the basic hydrogen atoms within the universe. These "true" and "beautiful" quarks we now call "top" and "bottom" quarks were at their highest energy level. They stepped down to "strange" and "charm" quarks by emitting electrons and positions. The "strange" and "charm" quarks stepped down to "up" and "down" quarks by emitting another series of electrons and positions as they expanded out into space and away from each other.

The inner expansion of space and time deformed the perfect sphere of particles that exploded out into the expanding space forming enormous stars and galaxies. The emission of neutrinos with fusion formed the heavier elements in the universe and locked out the reversal of events that created the universe as we know it.

One might ask what happened to all the anti-neutrinos that changed all the anti-particles to particles. They are still out there in the universe but with the expansion of space the anti-neutrinos have lost a great amount of their energy making it impossible for them to interact with existing quarks that have also stepped down from their higher energy levels. The anti-neutrinos will never be able to change existing quarks back to anti-quarks unless all their lost energy is returned.

Chapter Two . . . Space and Time

When Albert Einstein stated the presence of mass curved space and time around it, he gave space and time a property. Space and time could be changed which made space-time an entity equal to force, energy, and mass within the universe. One dictionary defined space as an empty void within which all other entities exist. Space is hardly an empty void. It is the last frontier to be understood by physicists.

The amount of knowledge we currently have about space could be written on the head of pin. Understanding the true nature of space will be our last major break throughs in physics. There is one profound statement that will help guide us to understanding the universe. "The universe is the way it is because it is the only way it can be."

This means our theories about the universe cannot contradict with what the universe does. It also means the universe is uniform. When our theories describe what happens in the universe then our theories stand true. It also means we cannot have two or more theories that contradict each other. Only one or none of the theories stand true.

For us to understand the properties of space, we must start at the beginning when the universe first created itself. This means space and time had to be created before energy and mass were created. There are only three entities that exist within the space of the universe, force, energy, and mass. We must have a clear understanding of what each entity is:

Force is the ability to cause change.

Energy is the amount of change occurring.

Mass is the resistance to change.

In the beginning, only Force existed. This Force was the universe. Force expanded and its value diminished, creating space and time. For us to understand force, we must have a system of measurement and use classical examples so our brains can visualize the existence of Force since we cannot see it. The standard system of measurement I shall use is grams, centimeters, and seconds. There are twenty-eight grams to an ounce. A centimeter is slightly less than a half of an inch. There are approximately thirty centimeters to a foot. A second is only a brief moment in time. The classical example I shall use to represent Force is a circle or a dot.

Power of Force

I want you to imagine a circle with a radius of 6.62×10^{-27} of a centimeter. Take a centimeter or slightly less than a half of an inch and divide it into ten equal parts. Take one of the ten parts and you now have 10^{-26} centimeters to go. Take one part and divide it into ten parts again and do this, twenty-six more times. What you end up with is a space smaller than the hydrogen atom, smaller than the proton within the hydrogen atom, smaller than the three quarks within the proton, smaller than the gluons that bind the quarks together, and a trillion times smaller than one tenth of a gluon. It is the curvature of this small circle that shall represent the power of Force. Force is the ability to cause change. The curvature of this circle is changing at every point around the circle.

Now I want you to imagine this circle as a sphere of Force. This sphere of force now has three dimensions with a radius that represents space. Time is our fourth dimension. Using our standard of measurement, we can determine the power of the Force when the universe began having a size or radius of 6.62 x 10^{-27} of a centimeter. We have a basic equation that can determine the power of Force:

$$\text{Force} = \text{Mass} \times \text{Acceleration}$$

$$F = MA$$

$$M = E / C^2$$

$$A = V^2 / R$$

$$F = (E / C^2) \times (V^2 / R)$$

$$F = E / R$$

$$F = 9 \times 10^{20} / 6.62 \times 10^{-27} = 1.3595166 \times 10^{47} \text{ over the power of one}$$

    The value of C is the speed of light. Nothing can travel through space any faster than the speed of light. The value of C is one of only two constants within the universe. The other constant is h or Planck's constant for energy. These two constants govern the entire universe. Acceleration is only the change of direction which is an increase in velocity in that direction. If one was traveling around our circle with a radius of 6.62 x 10^{-27} centimeters, one's velocity would be equal to the value of C^2. So, in our equations above, V^2 or velocity is equal to the value of C^2. They cancel each other out leaving us with the equation above where F = E / R.

    The amount of energy within a gram of mass is 9 x 10^{20} units of energy. The radius of our circle or sphere of force is 6.62 x 10^{-27} of a centimeter. The amount of Force it took to create the universe was 1.3595166 x 10^{47} over the power of one. This is the amount of Force our circle or sphere of Force had at the moment of creation.

Expansion of the Force

This sphere of Force began to expand outward at a velocity of C or the speed of light. The value of our Force diminished as it expanded outward creating space and time. It was not until our Force diminished to a value of one when energy and mass were created. The radius of the universe was 9×10^{20} centimeters. Strings of energy with a radius of h or 6.62×10^{-27} of a centimeter and a force value of one stepped down from the velocity of C and created particles with mass. The force of the universe continued to expand outward after it created all the mass particles within the universe.

The universe will continue to expand outward until it reaches a radius of 1.3595166×10^{47} centimeters and the value of force of the universe is equal to h or Planck's constant. At that time, the universe will come to an end and begin again with a radius of 6.62×10^{-27} of a centimeter and a force value of 1.3595166×10^{47} over the power of one. It will take 1.43994×10^{29} years before the universe comes to an end. The universe will end 143.994 billion-billion-billion years from now. We can easily imagine a billion years since our government now spends three and a half billion dollars a day. Expressing the end of the universe in billions of years does not seem like such a long time from now but it is as close to infinity as one can imagine. It will give us plenty of time to explore the universe if the human race does not come to an end before we find the means to explore the universe.

The Event Horizon is still expanding outward at the speed of light by the means of expanding the existing space already created. Space is expanding within itself to expand outward. However, this expansion within itself is spread out equally over the volume of space that exists today. So, the amount of inward expansion is very small over such an enormous amount of space. The moon is expanding away from the earth at three quarters of an inch per year. What is happening is the space between the earth and the moon is expanding three quarters of an inch per year. One might ask how this can be since we are in motion through space. Do not forget that all the space we are traversing is also expanding that three quarters of an inch every year. The greater the distance between two objects in space the greater the amount of expansion there is between the two objects. We will see this expansion of space as the two objects accelerating away from each other.

We will not be able to explore the universe using rockets as our means of propulsion. Nothing can travel through space any faster than the speed of light. The only way we will be able to explore the universe is to find a way to side step the space between us and our destination and travel faster than the speed of light by stepping down into the expanded dimensions of space. For us to do this, we must have a complete understanding of space and time.

Revision of Relativity

Albert Einstein gave us the theory of relativity. The only problem with his theories was they were incomplete. I will attempt to complete his theories of relativity by revising them just a little bit. Albert Einstein stated there was no absolute frame of reference with a zero velocity when it comes to determining the velocity of one through space. He stated all objects in space are in motion relative to each other which is true. However, there is an absolute frame of reference when it comes to determining one's velocity through space. The frame of reference is the speed of light.

When a rocket lifts off from the surface of earth, we say the rocket is accelerating away from us and its velocity is increasing towards the speed of light. This could not be further from the truth. The rocket is not accelerating towards the speed of light. It is accelerating away from the speed of light since it is gaining mass as it accelerates through space. Every particle that exists in the universe stepped down from the speed of light and this reduction of velocity was conserved in its mass. Remember, mass is the resistance to change. So, the more mass our rocket gains as it accelerates the more it is slowing down and away from the speed of light. Mass curves space and time and the more mass our rocket gains as it accelerates the more space and time curves around it. Time will slow down and space will curve so much our rocket's velocity through space will never equal the speed of light.

Velocity through Space

There is an equation one can use to determine our velocity through space relative to the speed of light. However, we cannot use the mass of just any particle. A proton and neutron are composed up of other particles we call quarks. The only particle that exists within the universe that is not composed up other particles is the electron. We must first convert the mass of the electron into units of energy. We do this by using Albert Einstein's famous equation:

$$E = MC^2$$

$$E = (9.109 \times 10^{-28}) \times (9 \times 10^{20})$$

$$E = 8.16 \times 10^{-7} \text{ units of energy}$$

The mass of the electron is equal to 8.16×10^{-7} energy. This amount of energy plays a key role in understanding the effects of gravity and the potential of space and time which we will get to later in this chapter. We will now use this amount of energy to determine our velocity through space relative to the speed of light.

$$V = \sqrt{E \times C^2}$$

$$V = \sqrt{(8.16 \times 10^{-7}) \times (9 \times 10^{20})} = 2.709 \times 10^7 \text{ centimeters per second}$$

The velocity of 2.709×10^7 centimeters per second is equal to 271 kilometers or 167 miles per second. It is also a velocity of 601,200 miles per hour through space. This is also the average velocity through space our galaxies have in the universe relative to each other. Creation of our mass has slowed our velocity down from 186,000 miles per second to 167 miles per second through space relative to the speed of light. This is quite a reduction in velocity which shows how much energy it takes to create a small amount of mass.

Now let us consider particles such as protons and neutrons. The quarks within them have been slowed down even greater amounts relative to the speed of light for them to remain in such tiny orbits around each other within the proton and neutron. This is the reason protons and neutrons have masses 1,867 times greater than the electron.

$$G_v = (hv)^2$$

$$G_v = (8.16 \times 10^{-7})^2 = 6.67 \times 10^{-13}$$

I would like to point out an equation I developed that proves the gravitational constant of the universe is equal to the energy of the electron's mass times itself. The gravitational constant of the universe is equal to a 6.67×10^{-13} value. The electron is the key to understanding space and time within the universe. Also, the gravitational constant is not a constant. It is declining in value as the universe and space expands outward.

$$E = h\nu$$

$$E = (6.62 \times 10^{-27}) \times (1.233 \times 10^{20}) = 8.16 \times 10^{-7} \text{ value}$$

Many scientists use electron volts for measuring energy. I use the Planck's constant which is the smallest unit of energy in the universe. No matter how much energy you have you can always divide the energy by Planck's constant and come up with whole numbers. Planck's Constants do not come in half units. Many scientists state that the Planck's Constants is just a number. The hell it is. Planck's Constants is the basic unit of energy. Though the number of units might be enormous, one does not have to convert from electron volts to joules or some other unit of measurement which can be confusing.

Depths of Space

I want you to imagine the surface of a deep lake as the space we find ourselves in at this very moment. Now, I want you to imagine dropping a rock into the center of this lake. The rock penetrates the surface of the lake and waves are created that expand outward in all directions. These waves are first created by the displacement of the water by the rock within the lake. As these waves expand outward in all directions they become waves of compression. These waves are no longer displacement of the water but are the compression of the water molecules. They become waves of energy.

Now I want you to imagine the surface of this lake has no depth to it. When we drop a rock onto the surface of the lake it cannot penetrate the surface of the lake. The energy from the rock being dropped is returned to the rock and the rock bounces back from whence where it came. No waves are created on the surface of the lake because it has no depth to it.

The space we exist within now has a tremendous amount of depth to it which I call dimensions. Space had to have a specific amount of depth value or potential to it before strings of energy could be created and penetrate the depths of space. Without these depths or dimensions in space, the universe could not exist. These dimensions of space also have a potential of force equal to 8.16×10^{-7} over the power of Planck's constant which is 6.62×10^{-27} value. All we do is divide the potential of space existing today by Plank's constant to determine the amount of expansion left in space.

$$P = (8.16 \times 10^{-7}) / (6.62 \times 10^{-27}) = 1.23 \times 10^{20}$$

Whatever the size the universe is today, its unexpanded dimensions of space must expand out 1.233×10^{20} times larger than the size of the universe is today. The universe has already expanded out 1.237×10^{54} times greater than the size it was when the universe began which was 6.62×10^{-27} of a centimeter radius. In the beginning space had 2×10^{73} dimensions of unexpanded space. Not until the universe expands out to 1.3595166×10^{47} centimeters and the potential of space is equal 6.62×10^{-27} or Planck's constant will the universe stop expanding. At that time, the universe will end and begin again with a radius of 6.62×10^{-27} centimeters and a force value of 1.3595166×10^{47} over the power of one.

Electromagnetic Radiation

Electromagnetic Radiation or light is only fluctuations in this space potential. Imagine an electron as a fisherman's float attached to a line that floats on the surface of the water. When this float bobs up and down waves are created on the surface of the water. When an electron bobs up and down in this space potential electromagnetic waves are created but not in all directions. Light is polarized either up and down, forwards and backwards, or side to side.

Size of the Universe

So, what is the size of the universe today? That would depend on the age of the universe today. There exists a paradox when it comes to the size and age of the universe. At last report, scientists have seen light from a galaxy thirteen billion light years away from us. That would be 1.2272×10^{28} centimeters in distance away from us. Yet, the same scientists say the universe is only fourteen billion years old with their last estimate. If the universe started with a big bang how did these galaxies get so far away from us? They would have to be traveling at half the speed of light or 1.5×10^{10} centimeters per second to get that far out away from us. Our galaxy would have to be traveling at the same velocity in the opposite direction. I have already proven our velocity through space is 2.709×10^7 centimeters per second which is over 1,000 times slower than the speed of light.

One scientist resolved the problem by stating the space in the universe had a rapid inflation that flung matter out so far into space. It is a theory with many flaws. In the theory, it states infinite space existed before the big bang and all the mass within the universe existed within a space of 10^{-33} centimeters. The curvature of this small amount of space around this enormous amount of mass would have been so great it would have been a universal black hole with no chance of expanding outward into space. Mass particles had to wait to be created until enough space was created to avoid the universe being one big black hole or nonexistent.

Our universe expanded out from a radius of 6.62×10^{-27} of a centimeter to a radius of 9×10^{20} centimeters before strings of energy having a force value of one could penetrate the thickness or the potential of space and time. At a radius of 9 x

10^{20} centimeters, the value of our force diminished from a value of 1.3595166 x 10^{47} over the power of one to equal a force value of one. Space had a potential value of one at a radius of 9 x 10^{20} centimeters. Prior to that radius the force value was far too great and the expansion of space and time within the sphere of force was far too great for mass particles to exist. Strings of energy had to wait until the potential of space was equal to one before they could penetrate space. At that radius, all the particles that exist today were created. These strings of energy stepped down from the velocity of light and gained mass as they did so.

So, how do we determine the size and age of the universe? We must look to the electron again. Also, we must be able to collaborate our answers by more than one method. Space had a potential value of one at a radius of 9 x 10^{20} centimeters. The energy that created our particles at the event horizon was equal to one or 1.51 x 10^{26} units of "h" or Planck's constants. The rest mass of the electron is equal to 1.23 x 10^{20} units of "h" or Planck's constants.

$$\frac{1.51 \times 10^{26}}{1.23 \times 10^{20}} = 1.234 \times 10^6 \text{ units of "h" difference}$$

Now, potential of space was equal to the force value of one at the event horizon. If we divide the potential of one by the difference above we get the potential of space today:

$$1/1.234 \times 10^6 = 8.16 \times 10^{-7} \text{ potential of space today}$$

The potential of space has dropped from a value of one at the event horizon down to a value of 8.16 x 10^{-7} potential and as the universe continues to expand outwards the potential of space will continue to drop down to a value of 6.62 x 10^{-27} value or one Planck's constant at which time the universe will stop expanding and come to an end.

$$9 \times 10^{20} \times 1.234 \times 10^6 = 1.1 \times 10^{27} \text{ centimeters}$$

For us to find ourselves in space with a potential of 8.16×10^{-7} value means the universe has expanded 1.234×10^6 times larger than 9×10^{20} centimeters. We have travels 1.1×10^{27} centimeters out in space from the point of the event horizon or 9×10^{20} centimeters.

Now that we know how far we have traveled in space we must determine what our average velocity through space has been. Then we can determine how long it took us to travel out to 1.1×10^{27} centimeters in space

$$V_{(average)} = \sqrt{(V_1 \times V_2)} = \sqrt{(3 \times 10^{10}) \times (2.709 \times 10^7)}$$

$$\text{Velocity}_{(average)} = 9.015 \times 10^8 \text{ centimeters per second}$$

Our particles created at the event horizon stepped down from the velocity of light or 3×10^{10} centimeters per second and we have already determined our current velocity through space relative to the speed of light is 2.709×10^7 centimeters per second. Using the equation above our average velocity through space has been 9.015×10^8 centimeters per second.

$$3.14712 \times 10^7 \times 9.015 \times 10^8 = 2.837 \times 10^{16} \text{ cm per year}$$

There are 3.14712×10^7 seconds in a year and our average velocity through space is 9.015×10^8 centimeters per second. We have traveled 2.837×10^{16} centimeters per year.

$$1.1 \times 10^{27} / 2.837 \times 10^{16} = 38.77 \times 10^9 \text{ billion years}$$

It took us 38.77 billion years to travel out to a distance of 1.1×10^{27} centimeters with an average velocity of 9.015×10^8 centimeters per second. With the event horizon expanding at the velocity of "C" or 3×10^{10} centimeters per second or 9.44×10^{17} centimeters per year we need only multiply 38.77×10^9 by 9.44×10^{17} centimeters to determine how large the universe is at this moment.

$$9.44 \times 10^{17} \times 38.77 \times 10^9 = 3.66 \times 10^{28} \text{ centimeters}$$

The first question you should be asking is if the universe is 38.77 billion years old why are we not seeing light from stars and galaxies 38.77 billion light years away from us? The answer to that question is because there are no stars or galaxies 38.77 billion light years away from us.

The second question you should be asking is if we are seeing light from galaxies thirteen billion light years away from us or 1.2272×10^{28} centimeters away from us and we have only traveled 1.1×10^{27} centimeters out into space how did these galaxies get so far away from us? Even if these stars are on the opposite side of the universe (2.2×10^{27} cms) from us they are 5.5 times further out than they should be.

The answer lies in how the universe is expanding at the speed of light or 3×10^{10} centimeters per second. The universe is expanding at the speed of light but this expansion is spread out over the existing space in the universe. So, the greater the distance between galaxies the more the space between them expands. It will take a mathematician greater than me to calculate this expansion within existing space but I think we will find that this expansion has pushed the galaxy out 5.5 times further out than we have calculated.

Mass particles cannot travel near or at the speed of light without giving up all their rest mass to do so. Mass particles cannot be accelerated away from the speed of light to equal the speed of light through space. Space would contract and time would slow down so slow the velocity of light would never be achieved not to mention it would take all the energy in the universe to accelerate one to a velocity through space equal to the speed of light through space.

Take note if one gives up one's mass to travel near the speed of light, time also slows down. So, whether you gain mass or lose mass, time is going to slow down in either direction. Local time is the time you exist in based upon your velocity through space that is determined by the mass of our electrons. All mass particles that exist in the universe today stepped down from the speed of light and gained mass. The question is how long did it take our mass particles to slow down from the speed light? A second would be at least a thousand times slower near the speed of light, say 2.709×10^{10} centimeters per second, compared to our one second at our current velocity through space at 2.709×10^{7} centimeters per second.

Travel through Space

The universe began as a sphere of force and it is still expanding as a sphere of force. This means all the space that exists in the universe is curved, naturally curved. So, no entity that exists in the universe is traveling in a straight line, relative to itself and all other entities in the universe. We see this curvature as gravity. Space in the universe is not flat. If a scientist states space is flat, they should join the flat earth society.

Here exists our chance to travel to distant planets, stars, and galaxies. Stepping down into these curved expanded dimensions of space will give us the ability to travel to distant planets. I want you to imagine a large circle. I want you to imagine the circumference of the circle as the space we exist in today. For us to travel through space, you must travel around the circumference of the circle. Let's say every four degrees of your circumference is equal to the distance of one light year. It will take you ninety light years to travel completely around your large circle.

Figure 22

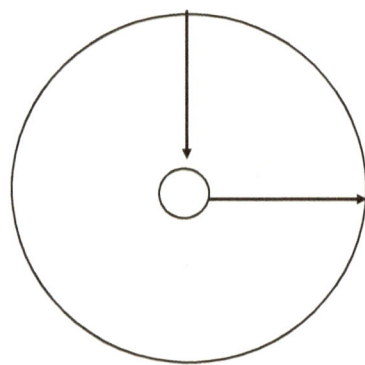

Now I want you to imagine a very small circle within your larger circle. Draw two lines from your larger circle to your smaller circle with ninety degrees between them as show in Figure 22 above. The distance between the two lines on your smaller circle will be much shorter than the distance on your larger circle. If you travel as fast around your smaller circle as you do on your larger circle you will reach the second line much quicker. If it takes you one light year to travel between the two lines on your smaller circle, you will have traveled twenty-two and a half years through space on your larger circle in one year. If your smaller circle is even smaller then you travel even greater distance in less time.

Stepping down into these curved expanded dimensions of space is the only way we can side step traveling through the space we exist in today. As long as dropping down into these expanded dimensions of space takes only a few seconds we can travel hundreds of light years in a few seconds, minutes, hours, or days. These expanded dimensions in space represent distance and not time. In a classical sense, time has stretched these expanded dimensions into distance. This will be the only way we are going to be able to travel to distant planets and stars within our life time.

If god does exist, he or she must have a means to travel from one end of the universe to the other almost instantaneously. It would be cruel and unusual punishment to create life in a universe so large as to deny that life the means to explore it. Fortunately, we do not have the means to do it yet. We still have too much religious baggage and hatred between ourselves to be spreading it out in the universe. This is why no one has come to contact us.

Puffs of Star Dust

As the universe expanded past a radius of 9×10^{20} centimeters it left behind a puff of star dust. Strings of energy stepped down from the velocity of light creating particles with mass. Particles and anti-particle having mass were created at the same time. There was a great implosion of the anti-particles. They emitted anti-neutrinos changing them into particles with mass. These particles, formed from strings of energy, divided four times creating the basic hydrogen atoms that would fuel the universe. Three quarks formed our neutrons and one of them emitted an electron changing the neutron into a proton giving us hydrogen.

This sphere of star dust was 18×10^{20} centimeters in diameter expanding outward at near the speed of light while stepping down from the speed of light giving us Mass = Energy / C x C or Einstein's famous equation $M = E / C^2$. This enormous puff of star dust gathered together to form immense galaxies that formed gigantic stars that created all the elements in the universe after they went super nova. New stars were born from the supernovas with planets around them. On some of these planets life was created. This life became intelligent and looked out into the universe and asked their god, "What the hell happened here?" Unfortunately, they never got their question answered by their god so they had to create science.

Infinities

Infinity is an age-old concept relating to an endless domain. For a millennium, scientists have pondered over its possibility and the implications regarding infinity. It seems to be a sound and logical concept for an individual to accept, yet for scientists it has been troublesome to deal with.

Infinities have plagued mathematicians and their equations from the moment they discovered pi extended forever, never to repeat itself or be brought to a final solution. Wherever an infinity raised its ugly head, progress slowed and the search for a solution ended in an uncertainty.

Space is our best-known example of an infinity. Albert Einstein dealt with space in his general theories of relativity and brought to light that the presence of mass and energy curved space. His theories did not prove whether space was finite or infinite. However, his equations did suggest the universe was expanding. Einstein went as far as introducing a constant into his equations to balance the known forces of the universe. He believed the universe was a static universe, neither expanding nor contracting.

Later, observations by a scientist name Edwin Hubble presented evidence the universe was indeed expanding at an alarming rate, suggesting an infinite space or universe. But if space was expanding and it was also infinite where was it expanding to and from. Einstein withdrew the constant from his equations and later admitted the introduction of the constant as the greatest blunder of his career. Now, we are trying to put it back in.

Modern scientists have suggested numerous theories to try and prove with enough mass in the universe, Einstein's space-time continuum might turn in upon itself and be truly finite. Unfortunately, the amount of mass they need is not present in the universe based upon current observations. They are missing about ninety per cent of the mass required for their equations. Now they are looking for dark matter. They call this dark matter W.I.M.P. short weak interacting mass particles. As of now, none has been found.

The point missed here by all and by Einstein is when Einstein suggested the presence of mass and energy curved space and time, Einstein gave this great void a physical property, it could be curved. By assigning space with a physical property, he made it a physical entity in the universe on par with force, mass, and energy. It could be altered or changed. Had space remained an unchanging void within which all force, mass, and energy resided, one could have automatically assumed space to be a true infinity. Existence of one infinity means other infinities could also exist.

The real problem with our concept of infinity is we must use physical properties to define it. In other words, if we find all physical properties of the universe are finite, our concept of infinity is invalid. Infinity becomes meaningless in the real physical sense. The question I like asking is did God give up infinity to create us?

Finite means limited. If one is to prove space to be finite, one must look for limits that exist in the universe. These limits must define finite in such a manner as to limit the physical universe to a specific size or dimension.

There are only two real limits or constants existing in the universe; (1) the power of C or the speed of light and (2) the power of h, units of change regarding energy. It is the relationship between C and h which will prove the universe to be either finite or infinite. How these two constants interact between each other will be the determining factors in our quest for an answer.

For one to comprehend the relationship between C and h, I will use classical examples to describe the powerful interactions between them. Classical examples are examples we experience in everyday life. It is better understood that a ball bouncing off a wall has reversed direction than to say a wave particle's wave function has collapsed with an elastic reaction. At the end of each classical example I will write out the cold and lifeless equations defining the mathematical relationship. Equations as accurate as they are cannot begin to describe the magnificent relationship they represent. As beautiful as the universe is, it is the simplicity of the universe that makes it far more beautiful. Equations tend to complicate and hide this beauty.

To understand the relationship between C and h, we must first have a basic understanding of these properties and other properties associated with them. I will introduce properties as we need them to keep the relationships simple. In many cases, the definitions of certain properties have been revised. Therefore, one who has a general knowledge of physics should not skip over these revised definitions as one will lose the essence of their relationship to other properties. The universe is the way it is because it is the only way it can be. If our current concepts of the universe do not fit our observations, we must change our concepts. One must also have a clear understanding of the terminology before one can comprehend the way the universe functions.

The Power of C

I use the term, power of C as it represents more than a specific velocity associated with light or electromagnetic radiation. It is one of two true constants existing in the universe. It represents a value and a limit that will always be true. It is an invariant. We will not find the power of C to be greater or less than the value it is in space.

$$3 \times 10^{10} = 30,000,000,000 \text{ centimeters per second}$$

We must first separate the power of C from light. Light or photons travel at the velocity of C which is 3×10^{10} centimeters per second. The power of C is the limit of velocity through the universe. There are other entities that also travel at the velocity of C. Therefore, C is not restricted or inherent just to light. Light is our best-known example of an entity that travels at C, so I will use light to help us understand and define the power of C.

The unique feature of light or photons is when they are emitted photons do not start out at a zero velocity from their source and over a specific distance and time reach the velocity of C. Photons are emitted at the velocity of C. When photons are absorbed or transmit their energy to a source their velocity does not decrease to zero before being absorbed.

Two scientists, Albert Michelson and Edward Morley, performed an experiment with light that proved the velocity of light was always constant in all directions regardless of the velocity of its source or the velocity of an observer who received the light. This discovery seems innocent enough on the surface but it represents a deep and profound problem to scientist and defied all known logic at the time. Spock would have broken down and cried had it not been for Albert Einstein and his special theory of relativity.

Here was the problem. If a person is in an automobile traveling along a road at a velocity of 10,000 miles per second (V_1) and turns on their headlights the light emitted from the headlights moved out away from the automobile at 186,000 miles per second (V_2). If you add the two velocities together

$(V_1) + (V_2) = (V_3)$ the total velocity of light should be 196,000 miles per second or (V_3).

Now a psychotic believes $2 + 2 = 3$. A neurotic knows $2 + 2 = 4$ but its makes them nervous. A mathematician can prove $2 + 2 = 4$. Yet, when it came to the total velocity of our example, the psychotic turned out to be right, $(V_1) + (V_2) = (V_2)$. A person standing at the end of the road who we will call the observer measures the velocity of the incoming light being emitted from the automobile's headlights and discovers it has only a velocity of 186,000 miles per second. The observer now takes out a radar gun and measures the velocity of the automobile. The observer discovers the automobile is indeed traveling at 10,000 miles per second towards them.

The simple math of adding two known velocities together failed. Mathematicians were in a crisis and the psychotics of the world were rejoicing. Fortunately, Albert Einstein came to the rescue of the mathematicians. The problem proved to be in the values used in their mathematics.

Before Einstein, mathematicians and scientist believed time and distance were absolutes or invariant regardless of one's frame of reference or velocity. They were giving time and distance the status of constants whereby making velocity an infinity. Albert Einstein simply reversed the status of distance and time and made the velocity of light a constant while making distance and time variables based upon one's frame of reference or velocity through space. A frame of reference is an observer's place and time in the universe. Depending upon one's velocity, the distance and time of one's frame of reference takes on different values. As one's velocity increases, distance contracts and time slows down. This is what happens to space as mass moves through it. By adjusting for the differences in distance and time, mathematicians prevailed over the psychotics. Whether Albert Einstein was aware of the importance of what he had done to infinities, he did eliminate an infinity from the universe. However, Einstein was not the first scientist to slay an infinity.

The Power of h

I use the term power of h, as it represents more than the numerical value of h which is used in scientific equations. Planck's constant which h is also referred too, deals with units of energy which are quantized. Planck's constant is the smallest value energy comes in. It is like pennies are to American dollars. You must either have one penny or two pennies. You cannot have 1.5 pennies. There is no in between or smaller units. If you have a lump sum of energy, the lump sum can always be divided equally by the value of h, a 6.62×10^{-27} value using a standard of measurement of grams, centimeters, and seconds.

6.62×10^{-27} = .000,000,000,000,000,000,000,000,000,066,200

This value was discovered by a scientist name Max Planck. He was researching black body radiation. At the time of his discovery, everyone believed energy flowed evenly and uniformly in the form of waves instead of particles. According to existing equations at the time, energy within a container could take on an infinite value as long as you continued to add energy. The more energy within the container, the shorter the wavelengths would become. Based upon their calculations, energy could take on an infinite number of shorter wavelengths representing unlimited energy values. However, their experiments did not agree with their calculations derived from their equations.

Max Planck, who preceded Albert Einstein, slew the infinite dragon with a constant. His observations forced him to acknowledge to the scientific community that energy came in discrete values that could be divided equally by the value of h.

This was such a profound discovery even Max Planck had difficulty accepting the discovery. He spent many years of his life trying to disprove what he had discovered. Change does not come easy in the scientific community. It took fifteen years for the scientific community to accept Einstein's special theory of relativity.

Planck's discovery proved energy to be quantized. As important as the quantification of energy was, h also represented the quantification of change. Therefore, h represents the smallest amount of change that can occur in the universe. Change relates to a measurable difference. If change occurs in any system or frame of reference, this change will be whole units of h or it will not occur. Planck's constant can also be considered as the smallest unit of force causing change over distance. Whether we like it or not there is a Planck's distance in space and a Planck's unit of time in space.

For one to exist in the universe, the smallest size an entity can have is a radius 6.62×10^{-27} of a centimeter. This size is relative to itself and all other entities of a greater size. Like C, h is a true constant. It is an invariant. Since these two constants stop infinities from occurring in the universe, the power of h and C have unique and universal governing properties. They govern the entire universe.

Chapter Three . . . Unification of Force

One of the primary goals of modern physics has been to unite all the forces and properties in the universe into one grand theory. James Maxwell was the first scientist to unite two seemingly unrelated forces. In 1864, he published his research acknowledging to the world that a magnetic field was only an electrostatic field in motion. This was a brilliant accomplishment. The invariant constant between the two fields was C or the speed of light. Using only four equations, he described the interactions occurring between static charged particles and magnetic fields.

Albert Einstein was the next scientist to unite two unrelated properties, energy and mass. Energy converts to mass and mass converts to energy at an invariant rate of C^2. The invariant constant links the two properties together in the same manner as C links the static and magnetic fields.

I will unite the Strong, Charge, Color, Mass and the Weak Force using the two constants "C" and "h" that limits the entire universe. Simply stated, a universe without limits cannot exist. The speed of light is incredibly fast to us but compared to the vastness of the universe is it slow, at a snail's pace. We cannot continue to split space in half before we reach 6.62×10^{-27} centimeters that is the limit or barrier to space. Planck's constant is the smallest unit of energy. A Planck's second is the shortest amount of time and a Planck's distance is the smallest unit of space. No change can be smaller than the value of "h", no change can occur in less time than a Planck's second, and no change can occur over a shorter distance than 6.62×10^{-27} centimeters. At one end, you have the limit "C" and at the other end you have the limit "h". In between you have the universe.

We can view the Singular Universal Force as a manual transmission with four gears. The first gear is the "Strong Force" that converted string energy into rest mass. It is active over a very short distance of 5.54×10^{-22} centimeters and is the most powerful of all four gears. The second gear is the "Charge Force" whose force is determined by its displacement of space and fields that act over the potential of space, 8.16×10^{-7}. The third gear is the "Color Force" which is the weakened "Strong Force" and like the "Strong Force" it is active over a short distance of 1.6×10^{-13} of a centimeter. The fourth gear is the "Mass Force" equal to the energy contained within the rest mass of the particle that resist change and curves space-time over a long distance creating the effect of gravity.

An outsider sees the difference in speed as the driver shifts through the gears and not knowing what a manual transmission is says there are four different forces accelerating the vehicle when there is only one transmission with four internal gears or "faces" delivering the force to the wheels of the vehicle.

We will learn that the Nuclear Force and all the other forces are not actual "faces" of the SUF or Singular Universal Force. They are extensions from the other faces. We will begin by reviewing the amount of energy within one unit of SUF.

$$F = 1/h = 1.51 \times 10^{26} \text{ units in Planck's Constants}$$

The amount of energy within one force unit is 1.51×10^{26} units of Planck's Constants which is equal to 1.11×10^{-21} grams of mass as shown in the equations below. There are twenty-eight grams to one ounce. So, 1.11×10^{-21} grams is a very small amount of mass when compared to an ounce or gram but when you look to the mass of an electron which is 9.109×10^{-28} grams one force unit creates 1,218,575 more masses than an electron, 653 more masses than a proton, and is almost equal to the mass of three uranium atoms. That is a lot of energy and force available within the quantum world.

$$E = h\nu = (6.62 \times 10^{-27}) \times (1.51 \times 10^{26}) = 1$$

$$M = E/C^2 = 1/9 \times 10^{20} = .111 \times 10^{-20} = 1.11 \times 10^{-21} \text{ grams}$$

Now that we have an idea of the tremendous force we are dealing with we can understand how the SUF or Singular Universal Force of one could have created all the matter within the universal. I will show how all the forces below are simply different "faces" or the extension of "faces" of one SUF unit limited by constants "C" and "h".

Figure 23

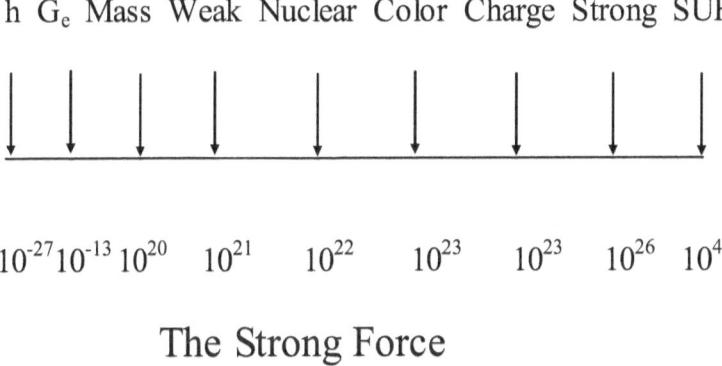

The Strong Force

The Strong Force is the opposing force to the Singular Universal Force and converts kinetic energy into mass which curves space giving us field structure we call particles with mass. As the Singular Universe Force diminished from a value of one to zero creating strings of energy that penetrated the space at the event horizon, the Strong Force emerged converting the string energy into rest mass as the SUF diminished to a value of zero. The Strong Force grew to a force value of one from the energy within the strings. Remember, you cannot lose what you begin with.

At the same time these strings of energy twisted in opposite directions creating the Charge Force equal to the Strong Force. The Strong Force and the Charge Force weakened becoming the Color Force that created quarks and gluons with three quantum states of charge, 1/3, 2/3, and 3/3. Quarks exchanged mesons and emitted W particles, electrons and positions, and formed protons and neutrons with three colored quarks held together by gluons. "C" and "h" determined the outcome:

$$S_F = 1/h = 1.51 \times 10^{26} \text{ units in Planck's Constants}$$

$$G_v = (hv)^2 = ((6.62 \times 10^{-27}) \times (7.55 \times 10^{25}))^2 = 2.498 \times 10^{-1}$$

$$E = hv = (6.62 \times 10^{-27}) \times (7.55 \times 10^{25}) = .49981$$

$$M = E/C^2 = .49981 / 9 \times 10^{20} = .0555 \times 10^{-20} = 5.55 \times 10^{-22}$$

$$F = G_v \times M_1 \times M_2 / R^2 \quad \text{and} \quad R = \sqrt{G_v \times M_1 \times M_2 / F}$$

$$R = \sqrt{(2.498 \times 10^{-1}) \times (5.55 \times 10^{-22}) \times (5.55 \times 10^{-22})} = 2.77 \times 10^{-22} \text{ cm.}$$

Time = 385,000 Planck seconds or 2.55×10^{-22} of a second.

The Color Force

The Color Force is an attractive and repulsive force existing within and between quarks within protons and neutrons that have three variant force values we define as Red, Green, and Blue. The color force started out with the color of "White" before it weakened to hold three quarks within the protons and neutrons. It was the Strong Force that converted all the string energy within the strings to mass that created the first Higgs Particles. The strength of the color force is a variant limited to the power of one which was the strength of SUF when mass particles were created at the edge of the event horizon. The equation below determines the force value of the color force and "h" is the constant factor in the equation:

$$\text{Color Force} = \frac{R^2}{h}$$

The three quarks within a proton or neutron can only be separated from each other by a center radius of 8.136×10^{-14} of a centimeter. At that radius, the color force of the quarks is equivalent to a value of one. With a color force value of one, all energy supplied to the quarks to separate them any further is converted to mass. It is equivalent to the Strong Force.

$$C_F = \frac{R^2}{h} = \frac{(8.136 \times 10^{-14})^2}{6.62 \times 10^{-27}} = \frac{6.62 \times 10^{-27}}{6.62 \times 10^{-27}} = 1$$

The radius of a proton and neutron just happens to be 8.136×10^{-14} of a centimeter. The color force of the quarks within a proton structure limits the size of a proton and neutron to a radius of 8.136×10^{-14} of a centimeter. The diameter of a proton and neutron is 1.6×10^{-13} of a centimeter or less.

Figure 24

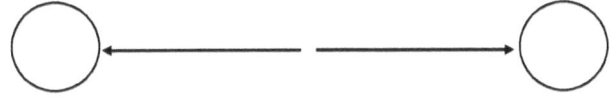

Since the color force has a limited radius range, one would not feel the color force from a quark or proton until one gets within 8.136×10^{-14} of a centimeter. At such a distance, watch out! It would be like falling off a cliff. You would suddenly accelerate towards the force. Let the Force be with you because you are going to need it to escape. Having fun, I must now correct the above statement. If an entity does not have an individual color or it is colorless, the entity will not feel the color force. An electron is a colorless or white free quark and if you shoot it into a proton structure it will not experience the color force.

The color force seems to have a strength value 295.6 times that of the charge force even though it is equivalent to the strength to the charge force. The reason for this is because of distance. Let's look at the force values of charge and color at a distance of 8.136×10^{-14} of a centimeter. The color force has a value of one resulting in a strength value of 1.09×10^{26}. The charge force has a strength value of 3.687×10^{23} or (3×10^{10} / 8.1366×10^{-14}).

$$\frac{\text{Color Force}}{\text{Charge Force}} = \frac{1.09 \times 10^{26}}{3.68 \times 10^{23}} = 295.6 \text{ times stronger}$$

Since the quarks within a proton structure have a rest mass about 300 times greater than an electron, we must divide 295.6 by 300 giving us the true observed force values or a strength value equal to the charge force. At 8.136×10^{-15} of a centimeter, any difference in our two force values balance out.

The diameter of a proton and neutron will fluctuate as will the strength of the individual colors. The quarks are not the only particles within a proton structure. There are gluons within the structure. We will learn their distance from the quarks at any given point will either weaken or strengthen the values of the color force. The nucleus of atoms is in a constant state of vibration or flux. We are not dealing with dead nuclei.

The Charge Force

We will now look at the values of charge and see if they are equivalent to our two other force values. When it comes to charge, we must first derive some values to use in our calculations. Charge of a particle is determined by its displacement of space (size) which is determined by its quantum energy in its rest mass. One coulomb of electrons (6.24×10^{13} electrons) will generate an electric field of one volt. In the equation below, F_n will represent force in Newton's, E_f will represent the electron's field strength, and q will represent coulomb's units. We first must derive the field strength of a single electron.

$$E_f = \frac{F_n}{q} = \frac{1}{6.24 \times 10^{13}} = 1.602 \times 10^{-14}$$

The field strength of an electron compared to one volt has a value of 1.602×10^{-14}. To get a force value within the field we must multiply the field strength value times itself and times a k_c value of 9×10^4.

$$F = \frac{k_c \times q^2}{R^2}$$

When it comes to mass, we derived our force value directly from the energy value representing the mass of the electron. With charge, we have the field strength value of 1.602×10^{-14} derived from the number of electrons in a coulomb field. The k_c value is like the gravitational value associated with mass.

$$\text{Force} = \frac{9 \times 10^4 \times (1.602 \times 10^{-14})^2}{R^2} = \frac{2.309 \times 10^{-23}}{R^2} = ?$$

Take note the values of $k_c \times q^2$ remains constant and it is the distance that determines the final values within the fields. So, let us see what our force value is at 2.722×10^{-17} centimeters?

$$\text{Force Value} = \frac{2.309 \times 10^{-23}}{7.409 \times 10^{-34}} = 3 \times 10^{10}$$

It is equivalent to the value of C at 2.722×10^{-17} centimeters where the field ends for the electron. The field strength of the electron is equal to the value of C. This is why the electron emits and absorbs photonic energy at the speed of light or C.

The Nuclear Force

Without nuclear binding, the universe would be a simple collection of hydrogen atoms. There would be no complex nuclei and elements. The color and charge forces with their force carrying particles perform an incredible balancing act binding the protons and neutrons together within a nucleus. Without the effect of gravity, protons and neutrons would have never had the chance to get close enough to bind. (The following nuclear bindings and examples are only theoretical.)

I will use two protons and one neutron to show how the two forces of color and charge perform this balancing act within the nucleus in Figure 25 below. It takes both forces and a little help from the weak force to bind and keep a nucleus together.

I have arranged the three structures to show maximum attraction. The charge of each quark is different resulting in attraction. The color of each quark is also different, adding to the attraction.

Figure 25

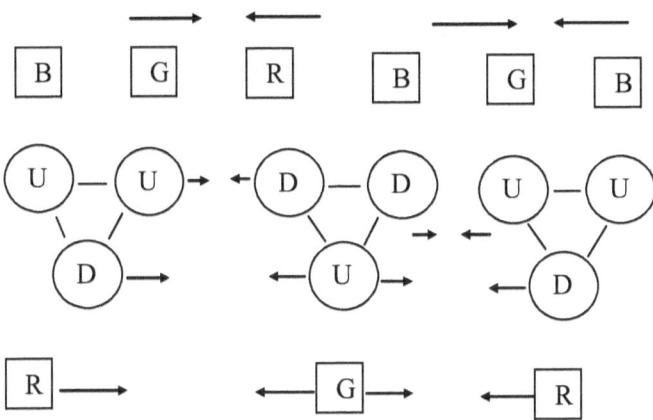

One must keep in mind it takes a great deal of energy to get two protons close enough so the color force can become effective and each quark can start to attract the other with the charge force. The presence of the neutron in the center dilutes the charge force existing between the two protons. This is why we find within a nucleus more neutrons than protons.

Figure 26

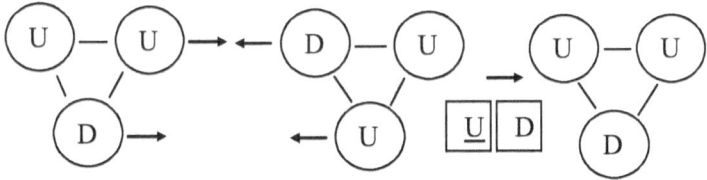

If the neutron in the center exchanges a −meson (minus meson) with the proton to the right, the neutron becomes a proton and the proton becomes a neutron. The exchange of the −meson is the weak force in action within the atoms. If we view the exchange of the −meson as a rotation of two structure's alignment, we see a state of maximum repulsion existing between the structures as shown in Figure 27 below.

The charge and color of each quark is aligned resulting in a maximum repulsion. The three structures would simply fly apart. Fortunately, the quarks within each proton structure is moving around and exchanging gluons within the structures, so there is never a state of fatal repulsion. ☺ I hope you did not miss my pun. One could even say there exists a state of neutrality between the forces if properly aligned as shown in Figure 28.

Figure 27

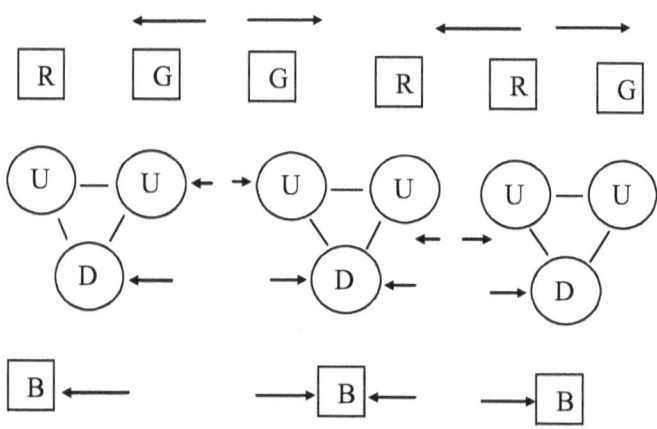

In Figure 28, the charge force is repulsing while the color force is attracting resulting in a reduction in the nuclear binding force between the structures. We are only dealing with two colors of the color in the examples. The binding force of two colors is 295.6 times stronger than the charge force. We know this because it is the two colors of a meson that binds the two quarks together even though their charge signs are the same. I have shown in the previous equations the binding force of two colors of the color force is about 295.6 times stronger than the charge force over the diameter of a nucleus but the rest mass of the quarks are 220 to 300 times greater. However, over the diameter of a nucleus the color force can be a hundred times greater than the charge force due to the neutrons in the nucleus. This is the reason the nuclei of the heaviest atoms are limited to about a hundred protons.

More protons would generate a combined charge force greater than the two-color force. The nucleus would divide into two or more nuclei. Nuclei with twenty-seven protons or more have a weaker nuclear binding force than nuclei with less than twenty-seven protons.

Figure 28

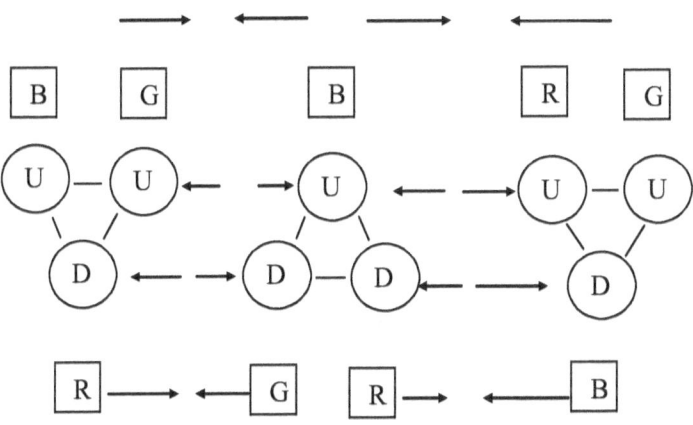

Helium with two protons and two neutrons form a very tight nuclear bond. We know this combination as the alpha particle. Viewing the three states of force arrangement there is insufficient attraction to keep the nucleus of the atom together especially when you add in neutrons and mesons over the nucleus' diameter of 6 x 10^{-13} centimeters. We must keep in mind a hundred protons reduce the nuclear binding force to a value of zero. Any collisions the nucleus under goes would be sufficient to tear the nucleus apart. This does not happen in the real world. So how do heavy nuclei stay together? It is simple.

We have over looked the most unique feature the color force has when we compare its strength value to the charge force.

The feature we have not yet considered is the third color of the color force as in Figure 29 below. Add the third color and conditions changes dramatically. The color force can bind three protons structures together as strong as its binds the quarks themselves within the proton structure. Once locked in by the color force, no force in the universe can separate them.

Figure 29

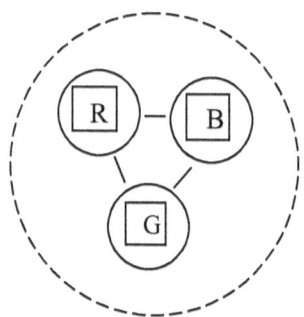

Drawing a circle with a radius of 8×10^{-14} of a centimeter, the circle represents the size of a proton (1.6×10^{-13} centimeters). Let us place three quarks at equal points within the circle. Placing a dot in the center of the circle represents the color white. Each of our quarks has its own color. If we move the three quarks towards the center of the circle, the three colors start to blend to form the color white. This means each individual color loses its strength. Let us reverse the situation and move all three quarks away from the center. Each individual color becomes stronger. The further away the colors are from each other, the stronger their force grows. Since all three colors are different, they all attract each other with a stronger and stronger force.

As the quarks move to the center, the color force weakens but the charge force becomes stronger over the shorter distance. The repulsion force becomes greater than the attractive color force. Moving the quarks away from center of the circle, the repulsion of charge becomes weaker while the attractive color force becomes stronger. Within the radius of a proton structure is where the two forces of color and charge come into balance.

Figure 30

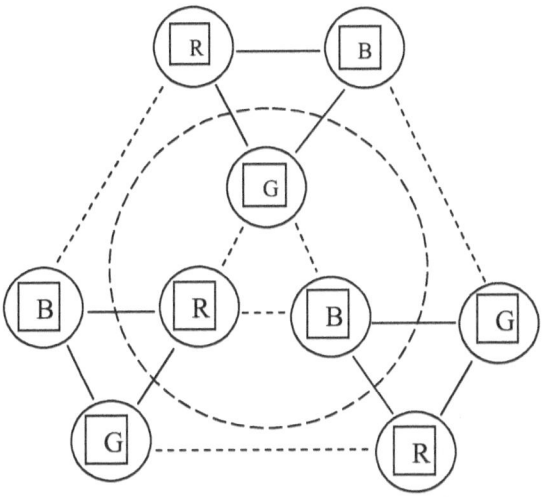

Let us take three proton structures and arrange them so one quark of each structure is within the circle as shown in Figure 30 above. Each quark within the circle must have a different color. This arrangement forms the same conditions we find inside of a proton structure. Color sees the distance half way between the three quarks as center point or as white. The conditions are no different than inside a proton structure. The three quarks are attracted with a binding force 295.6 times greater than the charge force. As the three proton structures are drawn together by the color force, the color force loses its binding strength.

The unique feature of color is when all three colors are present one cannot pull any of the proton structures out pass the circle. The three proton structures are forever bound within the circle no matter how much force is used to try and knock one out. If all the protons and neutrons form such a three-color relationship, the nucleus becomes an inseparable entity.

The color force binding between proton structures is not a permanent arrangement like it is within a proton structure of three quarks. Change one color of the quarks and the strong nuclear binding created by the different colors breaks down.

A helium alpha particle could form four of these strong nuclear bindings. The nucleus of atoms will be structured in such a way as to form as many of these strong nuclear bindings as possible. The so called strong force I call Nuclear Force is simply a color arrangement of protons and neutrons. Its strength values are only the difference between the changing and opposing force values of charge and color over distance. The Nuclear force does not actually exist as a real force in nature as some have expressed. It is only a difference in values.

The effect of gravity over 8×10^{-13} centimeters is too feeble to play an active role within the nucleus because the mass of the particles involved are still incredibly small. However, if the distance was reduced to a value of 2.72×10^{-16} of a centimeter (3,000 times smaller) between the quarks, gravity would take over. There would be one glob of quarks unable to interact. The individual properties of the quarks would begin to break down. Their mass would convert to energy and a black hole or mini event horizon would begin to take form. Fortunately, charge and spin keep this from happening. There are exceptions in the rare case where black holes do form due to extreme gravity.

The Weak Force

The difference between a Meson and W particle is the Meson changes a proton into neutron and a neutron into a proton within the nucleus without having to move them. The W and -W particle changes a proton or neutron by emitting an electron or positron changing the element of the nucleus. A uranium atom absorbing neutrons and emitting a -W particle or electron changes the uranium atom into a plutonium atom.

Figure 31

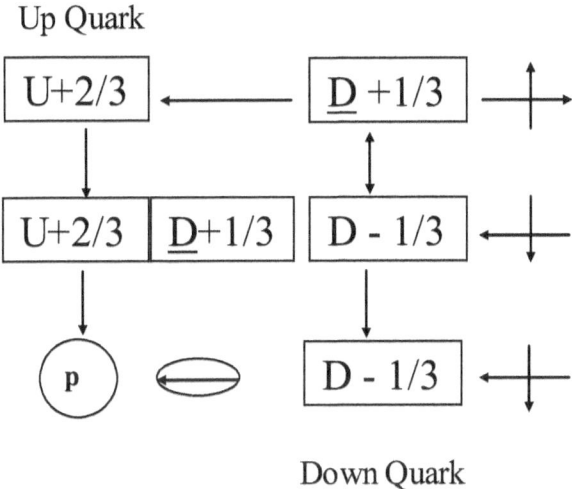

Down Quark

The above Figure 31 is an example of an Up quark changing into a Down quark changing a proton into a neutron by emitting a position and a neutrino. In Figure 32 below a Down quark is changing into a Up quark changing a neutron into a proton by emitting an electron and an anti-neutrino.

Figure 32

Down Quark

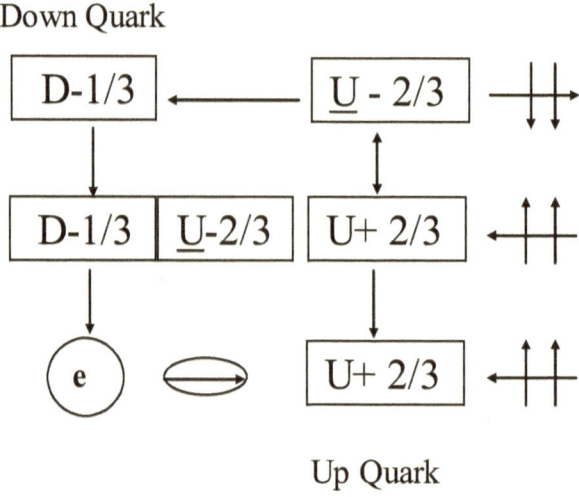

Up Quark

The Gravity Effect

Gravity is not a force where energy determines the force. Albert Einstein correctly stated the presence of mass contracts the space around the mass as it moves through space. Key point here is all particles having mass are in motion and as they move through a region of space that region of space contracts. In Figure 33 below we show a particle with mass moving into three regions of space. We see when the particle moves into region three the space contracts. As it moved from region three into region two the space contracts and the space in region three expands out to its original state. The direction of this expansion is in the same direction of its contraction and the effect would be space contracting. Finally, the particle moves into region one and region two expands back to its original state.

Figure 33

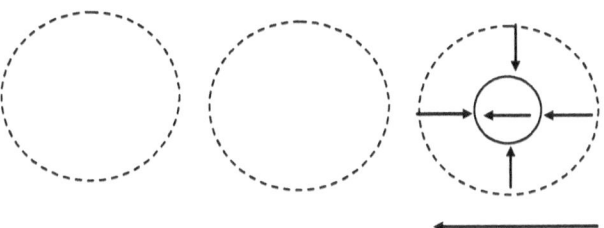

Figure 34

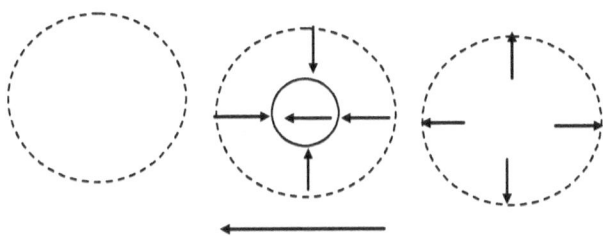

Figure 35

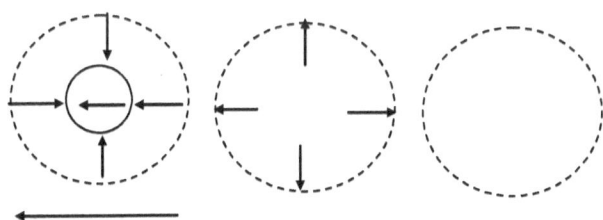

If gravity is an effect, then no energy is exchanged between two masses attracting each other. If gravity is an energy force, then energy will be exchanged between our two masses using inter-mediate force carrying particles such as gravitons. Gravitons have not been detected after decades of trying to detect them.

As space continues to contract around trillions upon trillions of mass particles their "charge" repels them until the gravity effect contracts the space between them enough for the "color" force to take over creating fusion within stars.

In Figure 36 below we have two particles moving into region three. The total sum of their mass contracts the space in region three. Our particles move together with this contraction of space. When they move into region two the space contracts and they move closer together with this contraction of the space between them.

Figure 36

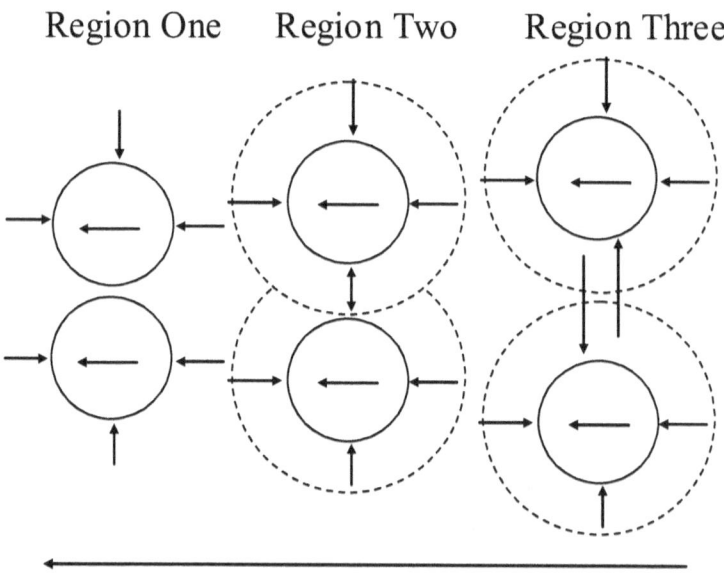

When they pass into region one the two particles are now almost touching each other. Each region of space represents a certain amount of time that has passed. Here is the important part to understand, no energy has been exchanged between them but the distance between them is now less giving us the gravity effect. Since the space curves between them their direction of momentum also changed towards each other.

The total mass of the two particles has contracted the space around them. We can say both particles have contracted into a lower level or dimension of curved space as shown in Figure 37 below. As they continued to travel together through space, distance and time continues to contract them together. Their total mass has contracted or curved the space around.

The scientific community now thinks of gravity as a force but it is only an effect created by the curvature or contraction of space by mass. As mass gathers together, it is reversing the expansion of space and contracting space and time back to its smaller dimensions. However, we can measure the strength of gravity by treating it as a force. We think of gravity as a weak force but as two particles having mass come closer together the strength of gravity grows greater as the distance between them becomes less. We will look at a Plank's distance.

$$F = G_v \times M_1 \times M_2 / R^2$$

$$F = \frac{6.67 \times 10^{-13} \times 9.109 \times 10^{-28} \times 9.109 \times 10^{-28}}{(6.62 \times 10^{-27})^2}$$

$$F = 1.263 \times 10^{-14}$$

We are going to use the mass of two electrons in the equation above to determine the strength of gravity over distance between two particles having mass. The gravitational constant is 6.67×10^{-13} but understand that the gravitational constant is not a constant. Its value is slowly declining with the expansion of space.

Figure 37

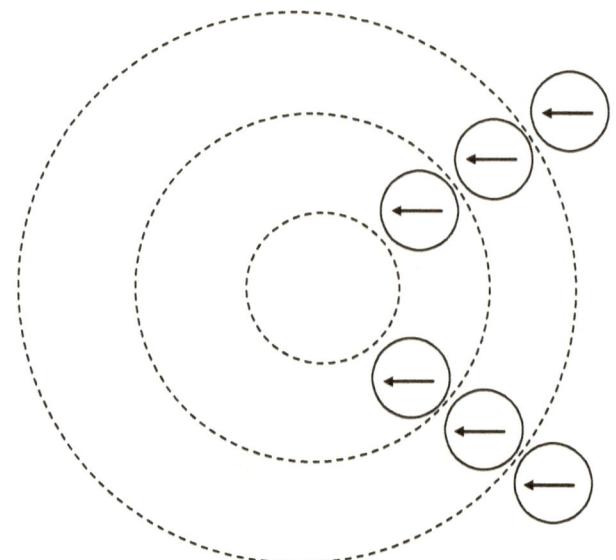

For us to get an idea of this gravitational strength above we can turn to the strength of the electron charge. One coulomb has 6.24×10^{13} electrons resulting one volt of electricity. The electric force of one electron can be determined by the equation below:

$$\text{Electric Force} = \frac{F}{q} = \frac{1}{6.24 \times 10^{13}} = 1.602 \times 10^{-14}$$

The gravitational strength between two electrons at a Plank's distance is almost equal to the charge of an electron.

Mass Force

No one ever thinks of mass as a force but the energy within particles having mass gives it a force equal to the energy units within the particle. Mass is the resistance to change. We incorrectly think a particle having mass takes up a certain amount of space within space. A particle having mass displaces space within space. The amount of energy the rest mass has determines the displacement or radius of the particle. Again, we must turn to the electron as it is the only particle that is not composed up of other particles.

First, we determine how much energy the rest mass of an electron has. We use Albert Einstein's famous equation below to determine the energy within the mass of an electron:

$$E = MC^2$$

$$E = 9.109 \times 10^{-28} \times 9 \times 10^{20} = 8.16 \times 10^{-7}$$

The radius of the electron can be determined by the equation below:

$$\text{Electron Radius} = \frac{hv}{C} = \frac{8.16 \times 10^{-7}}{3 \times 10^{10}} = 2.722 \times 10^{-17} \text{ cm.}$$

In Figure 38 below shows us the radius of an electron and its field strength at its radius. The field strength at the radius of an electron is equal to C and the force value of one diminished creating mass at the event horizon.

Figure 38

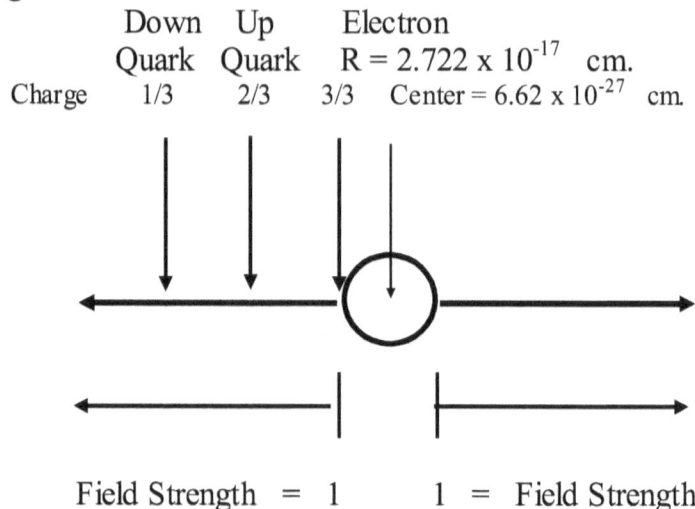

The gravitational value of a single particle is equal to its energy. The force value of the electron is equal to its energy times its mass. We can see the force value of the electron at its radius with the equation below:

$$F = \frac{E \times M}{R^2} = \frac{8.16 \times 10^{-7} \times 9.109 \times 10^{-28}}{(2.722 \times 10^{-17})^2} = \frac{7.409 \times 10^{-34}}{7.409 \times 10^{-34}} = 1$$

The force value of the electron at its radius is equal to the force value that created mass particles (minus an error factor of three hundredth times ten to the minus thirty fourth of a centimeter). We can also determine the field strength of the electron charge at the radius of the electron.

$$F = \frac{k_c \times q^2}{R^2} = \frac{9 \times 10^4 \times (1.602 \times 10^{-14})^2}{(2.722 \times 10^{-17})^2} = 3 \times 10^{10}$$

The purpose of all these equations is to show you how the constants "C" and "h" limits the entire universe and they cannot be exceeded. The equations also show what you began with you still have even though the "face" values of all our forces appear to be different.

One of the greatest phenomenon which I have not explained is how "charge" and mass joins together and creates fields of strength greater than the individual "charge" and mass of a singular entity. How does 6.24×10^{13} electrons with an individual charge of 1.602×10^{-14} of a volt combine their charges together to create one volt of electricity? How do trillions upon trillions of grams of matter combine to create a gravitational field that accelerates a gram of matter nine meters per second per second towards the center of the earth at sea level?

Notations:

Chapter Four . . . Equations and Calculations

Calculating the true size of the universe when it ends is a simple step in mathematics once you have a clear understanding of the two constants, C and h, that govern the entire universe.

Figure 39

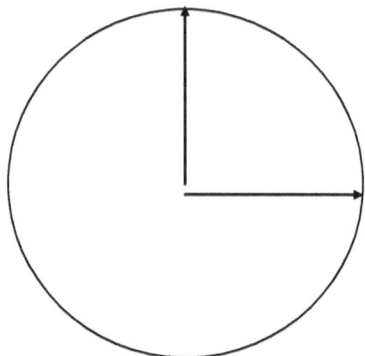

All we need is a means to calculate the radius of the universe when the SUF or Singular Universal Force diminishes to a force value of 6.62×10^{-27} or Planck's constant. Fortunately, we already have an established scientific equation which allows us to do this calculation.

$$\text{Radius} = \frac{\text{Velocity}^2}{\text{Acceleration}} \quad \text{or} \quad R = \frac{V^2}{A}$$

$$R = \frac{(3 \times 10^{10})^2}{6.62 \times 10^{-27}} = \frac{9 \times 10^{20}}{6.62 \times 10^{-27}} = 1.3595166 \times 10^{47} \text{ cm.}$$

All we do is put in the right numbers. The maximum velocity of the universe is C or 3×10^{10} centimeters. The minimum curvature of space over a distance of C^2 or 9×10^{20} centimeters will be h, Planck's constant, or 6.62×10^{-27} force value. We simply replace V^2 with C^2 and A with h. The radius of the universe when it ends will be C^2 / h or 1.3595166×10^{47} centimeters.

Figure 40

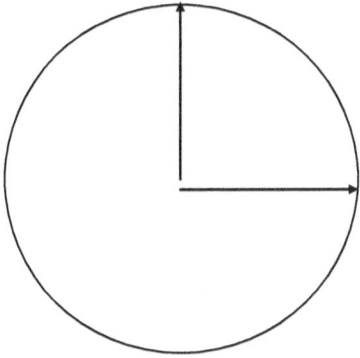

Expressing the radius of the universe in light years, the radius of the universe will be 1.43994×10^{29} light years. This is also the amount of time scientist estimate the proton will exist before it decays into a position and annihilates all the electrons in the universe. However, at 1.43994×10^{29} years from now the only things that will exist in the universe will be neutrinos and anti-neutrinos canceling out each other alone the event horizon of the universe.

Radius = 1.3595166×10^{47} centimeters.

Diameter = 2.7190332×10^{47} centimeters

Circumference = 8.5377642×10^{47} centimeters

Radius in Light Years = 1.43994×10^{29}

Diameter in Light Years = 2.8799×10^{29}

We now have the final size of the universe based upon the values of C and h. Get ready for the kicker. The universe is not that large yet. We are still expanding outward to that size. The universe is only 40 billion years old by my best estimates. This means the diameter of the universe is only 80 billion light years across. We still have 1.43994×10^{29} years to go until its final expansion size is reached.

```
 143,995,000,000,000,000,000,000,000,000
-                         40,000,000,000
 143,994,999,999,999,999,960,000,000,000  years
```
to go.

We still have an almost inconceivable amount of time remaining before the universe reaches its final expansion. This is as close to infinity as one can imagine. Yet, it is finite. Since we have the two master keys to the universe, let's unlock some more doors with them for the fun of it. By using established equations in the scientific community, we can find out how much mass, energy, and force comprises the entire universe. First, we will calculate the amount of mass the universe represents by using two equations.

$$F = M_1 A \qquad F = \frac{G_v M_1 M_2}{R^2}$$

We can combine the two equations since F is on the left side of both equations.

$$F = M_1 A = \frac{G_v M_1 M_2}{R^2}$$

We can drop the F from the equation since we are not looking for force. Since M_1 is on both side of the equation, we can cancel out M_1 by dividing both side of the equation by M_1.

$$A = \frac{GM}{R^2}$$

We multiply both sides of the equation by R^2 and divide both sides by G to get our final equation for mass:

$$M = \frac{AR^2}{G}$$

G is a gravitational value equal to 6.67×10^{-13} using our standard of measurements. We have already seen how this value is equal to the energy of the electron times itself. We have the values for R and A in our equation:

$$M = \frac{(6.62 \times 10^{-27}) \times (1.359 \times 10^{47})^2}{6.67 \times 10^{-13}} = 1.8 \times 10^{80} \text{ grams}$$

This value does not represent the amount of matter currently in the universe. It represents the entire universe as one mass. It includes all the existing matter, expended energy, and yes, the kitchen sink. Converting mass to energy, we use Einstein's famous equation:

$$E = MC^2 = (1.8 \times 10^{80}) \times (3 \times 10^{10})^2 = 1.62 \times 10^{101} \text{ energy}$$

We can convert the energy units to units of h since there are 1.509×10^{26} units of h per unit of energy (ergs.):

$$h = (1.509 \times 10^{26}) \times (1.62 \times 10^{101}) = 2.4445 \times 10^{127} \text{ units of } h$$

We have already calculated the value of the SUF or force that created the universe:

$$F = 1.359 \times 10^{47}$$

We can think of Force as magnitudes or power of 10 between the minimum force value of h and the maximum force value of SUF or Singular Universal Force. In other words, all the known forces existing in the universe must lie between the value of 6.62×10^{-27} and 1.3595166×10^{47} value. We will never find a force greater than 1.3595166×10^{47} over one or less than h or 6.62×10^{-27} value.

All the forces known to science can be marked along a force spectrum line showing their relative magnitude strength and position with regards to h and SUF show below in Figure 41.

Figure 41

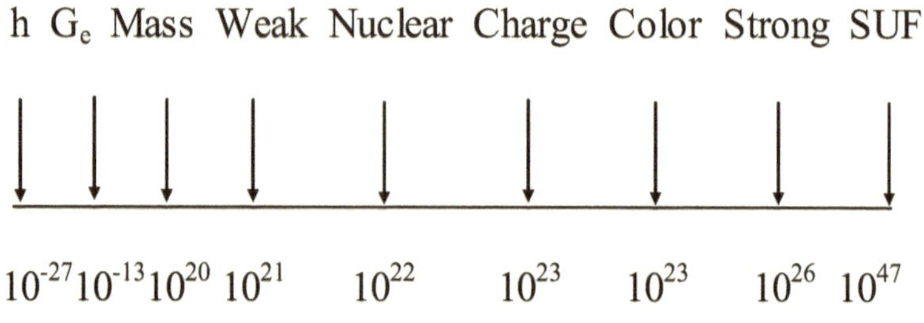

h G_e Mass Weak Nuclear Charge Color Strong SUF

10^{-27} 10^{-13} 10^{20} 10^{21} 10^{22} 10^{23} 10^{23} 10^{26} 10^{47}

h or Planck's Constant ... Minimum force and energy unit

G_e or Gravity Effect ... Contraction of Space by Mass

M or Mass ... Energy force within mass resistant to change

W⁻ minus and W⁺ plus ... Weak electron and positron transitions

N_F or Nuclear Force ... Force binding the nucleus together

C_F or Charge Force ... Electric charge force

C_L or Color Force ... Force binding quarks together

S_F or Strong Force ... Force converting energy to mass (Higgs)

SUF or Singular Universal Force/Force from all forces are derived

    There is a reason and a mathematical equation for each "face" value derived from our SUF or Singular Universal Force. The values above are relative to each other in strength but their true mathematical values are different. As we have seen the equations for each "face" deals with different distances between particles. The "Strong" force is only active within a radius distance of 6.62×10^{-27} and 2.77×10^{-22} centimeters having a strength of 10^{26} value. The "Strong" force diminishes into the "Color" force with a greater active distance with a radius greater than 2.77×10^{-22} and smaller than 8.136×10^{-14} of a centimeter having a strength of 10^{23} value. The "Charge" force diminished from a strength of 10^{26} value to 10^{23} value when four quarks decayed into three quarks within a radius smaller than 8.136×10^{-14} of a centimeter and greater than 2.77×10^{-22} of a centimeter. The "Weak" force is active over a distance greater than 8.136×10^{-14} of a centimeter emitting a W particle that decays into an electron that clouds around a proton. The greater the distance the weaker the force over the distance.

Chapter Five . . . Basic Particles 101

Quark Family

Bottom and Top

Figure 42

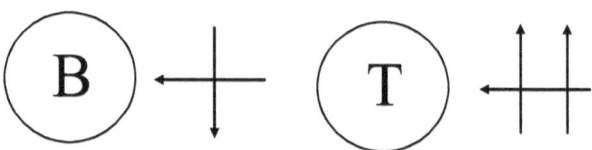

Strange and Charm

Figure 43

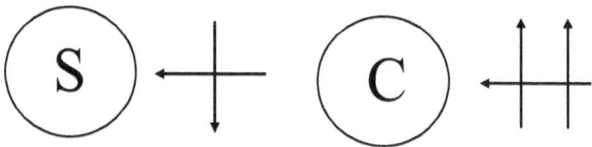

Down and Up

Figure 44

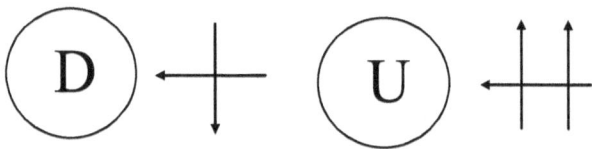

There are three levels of quarks each level being higher in energy than the other. They contain the same charge values even though they have higher energy levels. A down arrow represents a negative one third charge and an up arrow represents a positive one third charge. An arrow pointing to the left represents normal quark particles that exist in protons and neutrons.

Anti-Quark Family

Anti-Bottom and Anti-Top

Figure 45

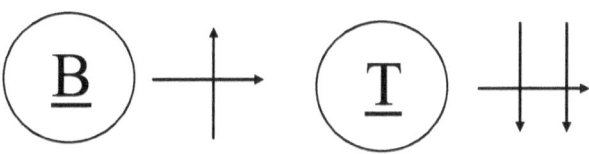

Anti-Strange and Anti-Charm

Figure 46

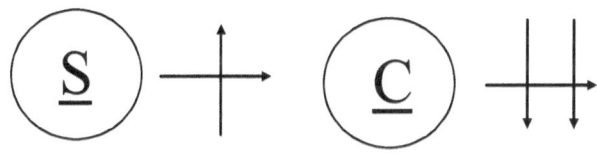

Anti-Down and Anti-Up

Figure 47

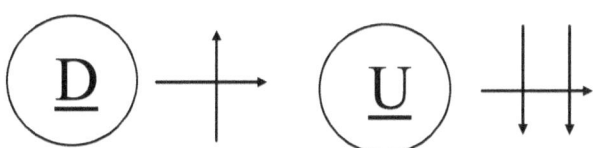

There are three levels of anti-quarks each level being higher in energy than the other. They contain the same charge values even though they have higher energy levels. A down arrow represents a negative one third charge and an up arrow represents a positive one third charge. An arrow pointing to the right represents anti-quark particles that do not exist except in anti-protons and anti-neutrons.

W Family

Figure 48

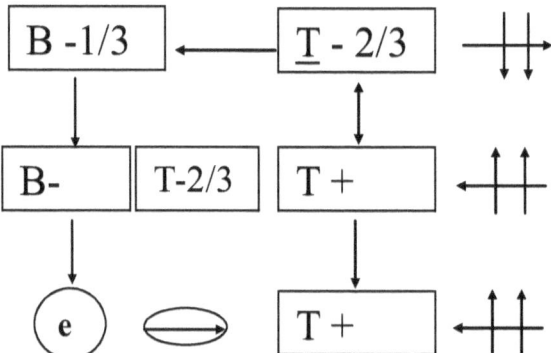

Figure 49

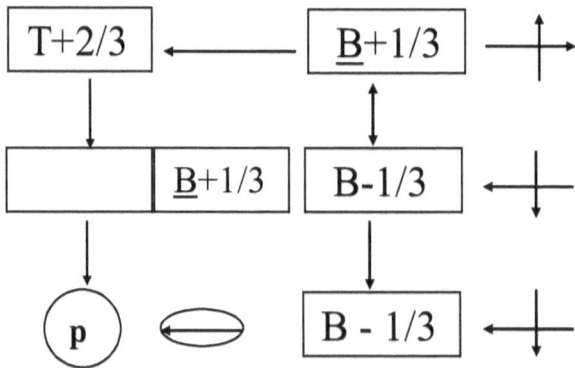

Figure 50

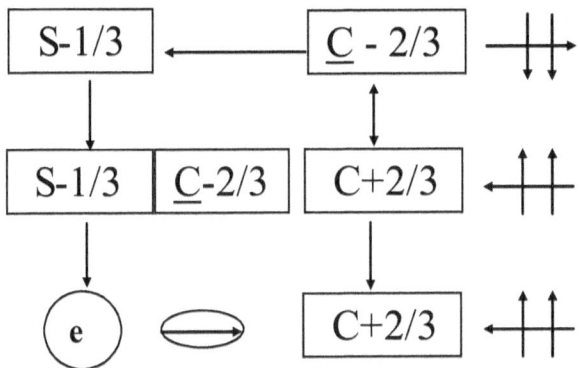

Figure 51

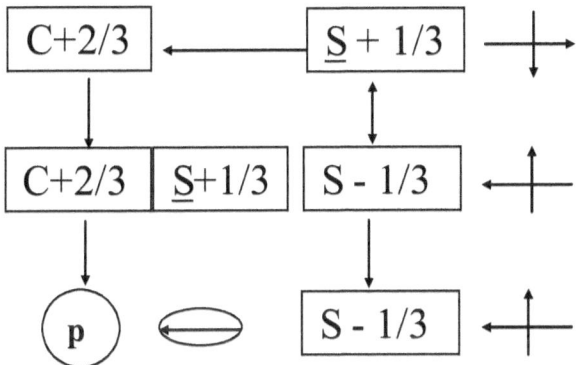

Figure 52

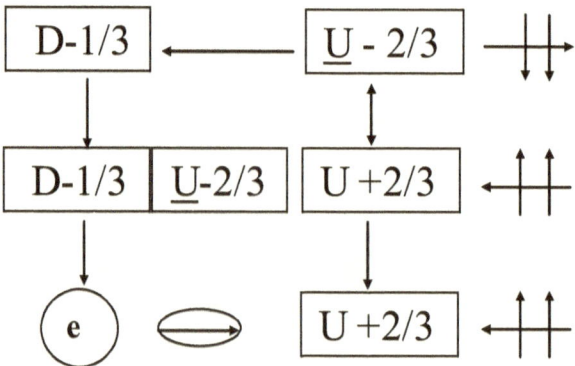

Figure 53

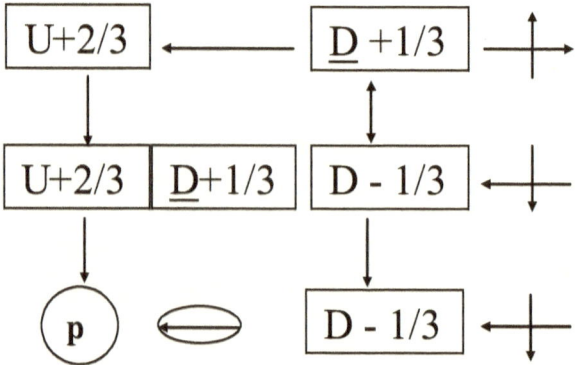

As we see all six W examples the upper left quark is replaced within a proton or neutron by the lower right quark by creating a pair of virtual Higgs particles and emitting the lower left pair of quarks that decay into an electron or positron and a neutral neutrino or anti-neutrino. All the quarks have color associated with them and the two colors in the W particles cancel each other out leaving the electron or positron colorless.

The –W and +W particles that cause the weak force are also the emitted particles that change a proton to a neutron and neutron to a proton. Meson are exchanged between protons and neutrons in the nucleus while the –W and +W are emitted from a nucleus to alter its element classification.

All the hundreds of particles we create in our accelerators are combinations of two or three quark systems forming mesons, electrons, protons, and neutrons. We cannot create a particle without creating its anti-particle. When energy divides, it creates a particle and anti-particle. All the anti-particles become particles by emitting anti-neutrinos or by annihilating its particle leaving two neutrinos, a negative and a positive neutrino.

There is no anti-matter existing naturally in the universe nor is there any unseen and undetectable dark matter existing in the universe. We are a universe of field particles.

Proton and Neutron Structures

Omega Particle

Figure 54

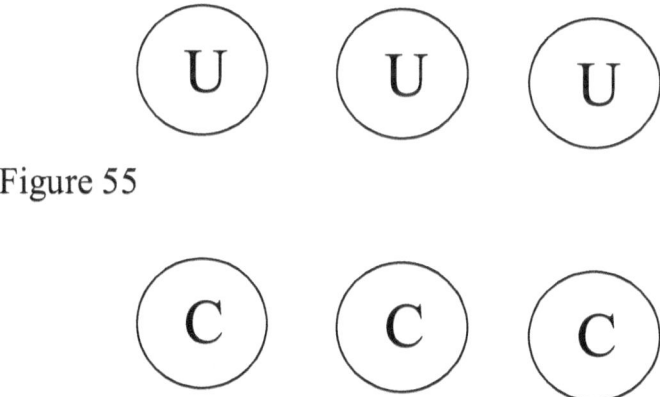

Figure 55

Figure 56

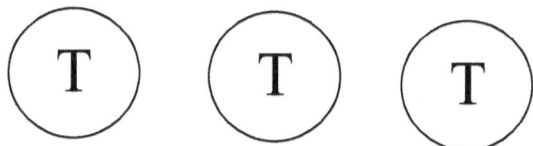

The Omega Particle has three Up quarks which give it a double positive charge. The higher states of the Omega Particle are three Charm quarks and three Top Quarks.

The Proton Structure

Figure 57

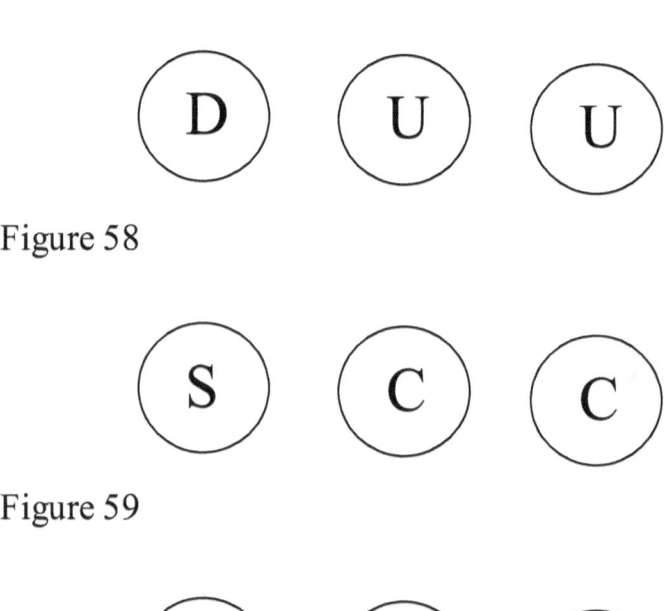

Figure 58

Figure 59

A proton consists of two Up Quarks and one Down Quark giving it a single positive charge. The higher states of the Proton are a Strange and two Charm quarks and a Bottom and two Top quarks.

Neutron Structure

Figure 60

Figure 61

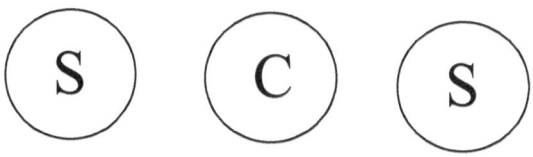

Figure 62

A neutron consists of two Down Quarks and one Up Quark giving it a neutral charge. The higher states of the neutron are two Strange and one Charm quark and two Bottom quarks and one Top quark.

Negative Proton Structure

Figure 63

Figure 64

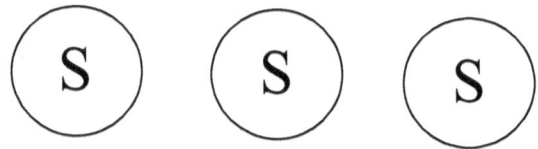

Figure 65

A negative proton consists of three Down Quarks giving it a negative charge. The higher states of the negative proton are three Strange quarks and three Bottom quarks.

Anti-Particle Structures

Figure 66

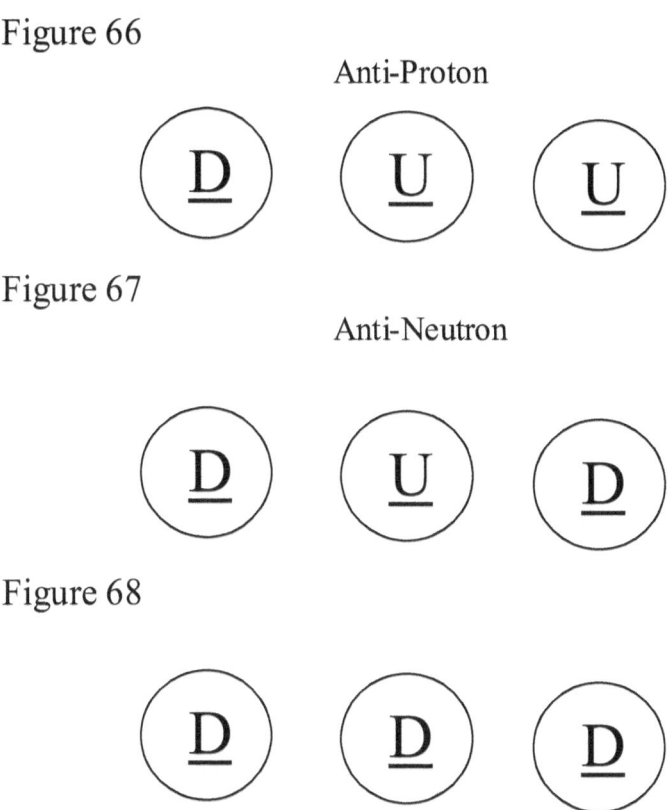

Anti-Proton

Figure 67

Anti-Neutron

Figure 68

There are anti-particle structures consisting of an anti-proton with a negative charge, anti-neutron, and an anti-proton with a positive charge. None of these particles exist naturally. We can create them in our accelerators given enough energy. They will be created in pairs with opposite charged particles.

Gluons

Only quarks emit gluons. Gluons are the force carrying particles representing the color force. There are three levels of quarks so it should not be a surprise to discover there are three levels of gluons. There are also three flavors associated with gluons. Since there are three quarks to a proton structure, there are three distinct differences between the quarks. The Pauli Exclusion Principle forbids two quarks from being the same or they must occupy the same space and time. The scientific community represents these differences by using the concept of individual colors. The three individual colors of red, green, and blue when combined form the neutral color white. The color white, not actually being an individual color, represents the neutral state of the color force.

The upper two levels of quarks do not exist naturally on earth. They have given up their higher mass levels and have become our common up and down quarks. One can assume what applies to one level of quarks also applies to the other levels. However, assumptions should not be considered as facts.

From this point onward, there is going to be a great deal of implied logic to define the force carrying particles we call gluons. A hypothesis involves implied logic since there is little experimental evidence to make a hypothesis a standing theory. All we have today is theory. They are theories because they are subject to change without notice based upon experimental evidence. Experimental evidence is fact. Facts can be proven by anyone who chooses to perform the experiment. It is the interpretation of the experimental facts that gives rise to so many theories. As long as there are differences between human beings, there is always a chance a fact can be interpreted in more than one way. This is the main reason why the search for truth and fact must never end.

Looking at the three quarks making up the neutron structure in Figure 69 below there are two D-quarks having the same energy level, the same charge, and the same direction of spin but they have different colors which allows them to coexist within a proton or neutron.

Figure 69

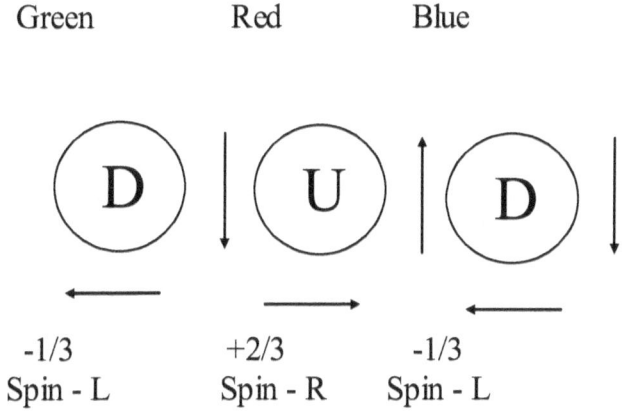

The strength of the color a quark has is not written in stone. The further the colored quark moves out or away from the other quarks within a proton or neutron the greater its color becomes. The closer they move together their color force diminishes. As long as there is difference of strength of the color force between the quarks they can coexist within a proton or neutron.

A "blue" quark can become a "red" quark by emitting a gluon. A "red" quark can become a "green" quark by absorbing a gluon. A "green" quark can become a "red" or "blue" quark by emitting or absorbing a gluon. Ranking the colors a "blue" quark has the highest energy level. The "green" quark is the intermediate color or strength. A "red" quark has the least energy level.

In a classical sense, we can say like colors repel each other while unlike colors attract each other and this attraction and repulsion is what keeps the quarks within the proton or neutron. What happens within a proton or neutron is when two unlike quarks come closer together their color force diminishes and when force diminishes energy is created in its place. This energy must divide creating a gluon which has two colors different than the color of the color force diminishing.

A "blue" quark whose color is diminishing as it moves to the center of a proton or neutron must create and emit a gluon with the colors "red" and "green" which is an anti-green. Emitting this gluon sends the quark outward from the center. This gluon is like the "Higgs" particles. Energy divides creating a particle and anti-particle. One of its colors must be an anti-color. Should a "green" quark absorb this "red and anti-green" gluon the anti-green in the gluon cancels out the green in the quark and the "green" quark becomes a "red" quark.

Our original "blue" quark must take on the color green since it created an anti-green color within the gluon it emitted. Colors must always balance out and anti-particles must always be eliminated from stable particle interactions.

There are six types of gluons quarks can emit depending on their color as show in Figure 70 below. When a "red" quark approaches a "green" quark the "red" quark's force must diminish enough to create a gluon. This is where we see two unlike quarks attracting each other because their color force has to diminish first. When a "red" quark finally emits a gluon, it must turn the color of the anti-color it created. For a moment in time our "red" quark turns "green" and we see two like quarks repulse each other while the gluon is in transit. The emission and absorption of the gluon sends our two quarks away from each other and their energy momentum is converted back to mass by the strong force.

Figure 70

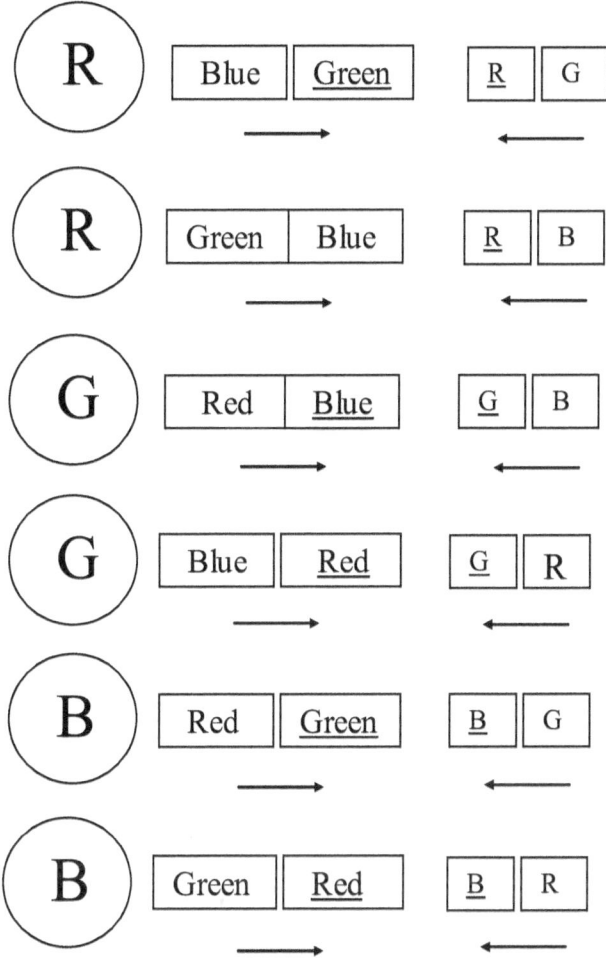

It is easier to view the interaction in a classical sense. When a "red" quark absorbs a "blue-green" gluon the "red" and "blue" colors cancel each other out and our "red" quark turns "green". One does not have to deal with the anti-colors the gluon contains. The same "red" quark can absorb a "green-blue" gluon and turn into a "blue" quark.

You are now at the end of this book and you should be at the beginning of your wonderment. Your studies in physics are just beginning. There are a million questions left unanswered in physics. Hopefully, you might contribute to answering some of the questions in the future.

THE END of this book is only the beginning of another book.
Will you be the one to write it?

www.ingramcontent.com/pod-product-compliance
Lightning Source LLC
Chambersburg PA
CBHW030859180526
45163CB00004B/1633